AQA KS3 Science

Student Book Part 1

Ed Walsh and Tracey Baxter

Pupil Referral Service
College Hall, Pupil Referral Unit
West Road, Off Old Wokingham Road
Wokingham, Berks RG40 3BT

William Collins' dream of knowledge for all began with the publication of his first book in 1819. A self-educated mill worker, he not only enriched millions of lives, but also founded a flourishing publishing house. Today, staying true to this spirit, Collins books are packed with inspiration, innovation and practical expertise. They place you at the centre of a world of possibility and give you exactly what you need to explore it.

Collins. Freedom to teach

An imprint of HarperCollins*Publishers*
The News Building
1 London Bridge Street
London
SE1 9GF

Browse the complete Collins catalogue at
www.collins.co.uk

© HarperCollins*Publishers* Limited 2017

10 9 8 7 6

ISBN 978-0-00-821528-6

All rights reserved. No part of this publication may be reproduced, stored in a retrieval system, or transmitted in any form or by any means, electronic, mechanical, photocopying, recording or otherwise, without the prior written permission of the Publisher or a licence permitting restricted copying in the United Kingdom issued by the Copyright Licensing Agency Ltd., 90 Tottenham Court Road, London W1T 4LP.

British Library Cataloguing in Publication Data
A Catalogue record for this publication is available from the British Library

Commissioned by Sarah Busby and Joanna Ramsay
Authors Tracey Baxter and Ed Walsh
Contributors to original material Sarah Askey, Sunetra Berry, Pat Dower and Anne Pilling
Copy edited by Jane Roth
Project managed by Siobhan Brown
Proofread by Helen Bleck
Cover design by We Are Laura
Cover images: arigato/Shutterstock, (tl) Artem Kovalenco/Shutterstock, (tr) fuyu liu/Shutterstock, (c) NikoNomad/Shutterstock, (bl) Pavel Vakhrushev/Shutterstock, (bc) robert_s/Shutterstock, (br) Sailorr/Shutterstock
Designed by Joerg Hartmannsgruber and Ken Vail Graphic Design Ltd
Illustrations by Ken Vail Graphic Design Ltd
Typesetting by Ken Vail Graphic Design Ltd

Printed and bound by Grafica Veneta S.p.A., Italy

Approval message from AQA

This textbook has been approved by AQA for use with our syllabus. This means that we have checked that it broadly covers the syllabus and we are satisfied with the overall quality. Full details of our approval process can be found on our website.

We approve textbooks because we know how important it is for teachers and students to have the right resources to support their teaching and learning. However, the publisher is ultimately responsible for the editorial control and quality of this book.

Please note that when teaching the AQA KS3 Science course, you must refer to AQA's syllabus as your definitive source of information. While this book has been written to match the syllabus, it cannot provide complete coverage of every aspect of the course.

A wide range of other useful resources can be found on the relevant subject pages of our website: aqa.org.uk

MIX
Paper from responsible sources
FSC
www.fsc.org
FSC C007454

FSC™ is a non-profit international organisation established to promote the responsible management of the world's forests. Products carrying the FSC label are independently certified to assure consumers that they come from forests that are managed to meet the social, economic and ecological needs of present and future generations, and other controlled sources.

Find out more about HarperCollins and the environment at
www.harpercollins.co.uk/green

HarperCollins
PUBLISHERS
Since 1817

Contents

How to use this book — 4

Chapter 1 Forces
Speed *and* Gravity — 6

Chapter 2 Electromagnets
Voltage and resistance *and* Current — 30

Chapter 3 Energy
Energy costs *and* Energy transfer — 52

Chapter 4 Waves
Sound *and* Light — 76

Chapter 5 Matter
Particle model *and* Separating mixtures — 102

Chapter 6 Reactions
Metals and non-metals *and* Acids and alkalis — 126

Chapter 7 Earth
Earth structure *and* Universe — 150

Chapter 8 Organisms
Movement *and* Cells — 174

Chapter 9 Ecosystems
Interdependence *and* Plant reproduction — 198

Chapter 10 Genes
Variation *and* Human reproduction — 220

Glossary — 242

Index — 252

How to use this book

This page covers ideas you have met before so you can check your understanding before learning about the next Big Idea.

This page summarises the exciting new Big Idea you will be learning about in the chapter.

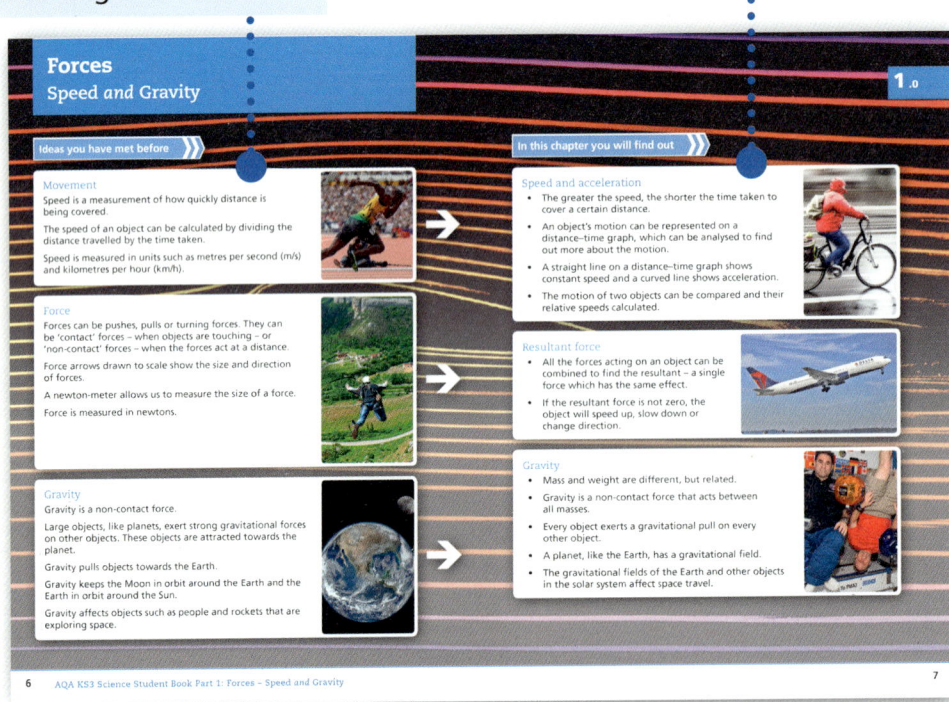

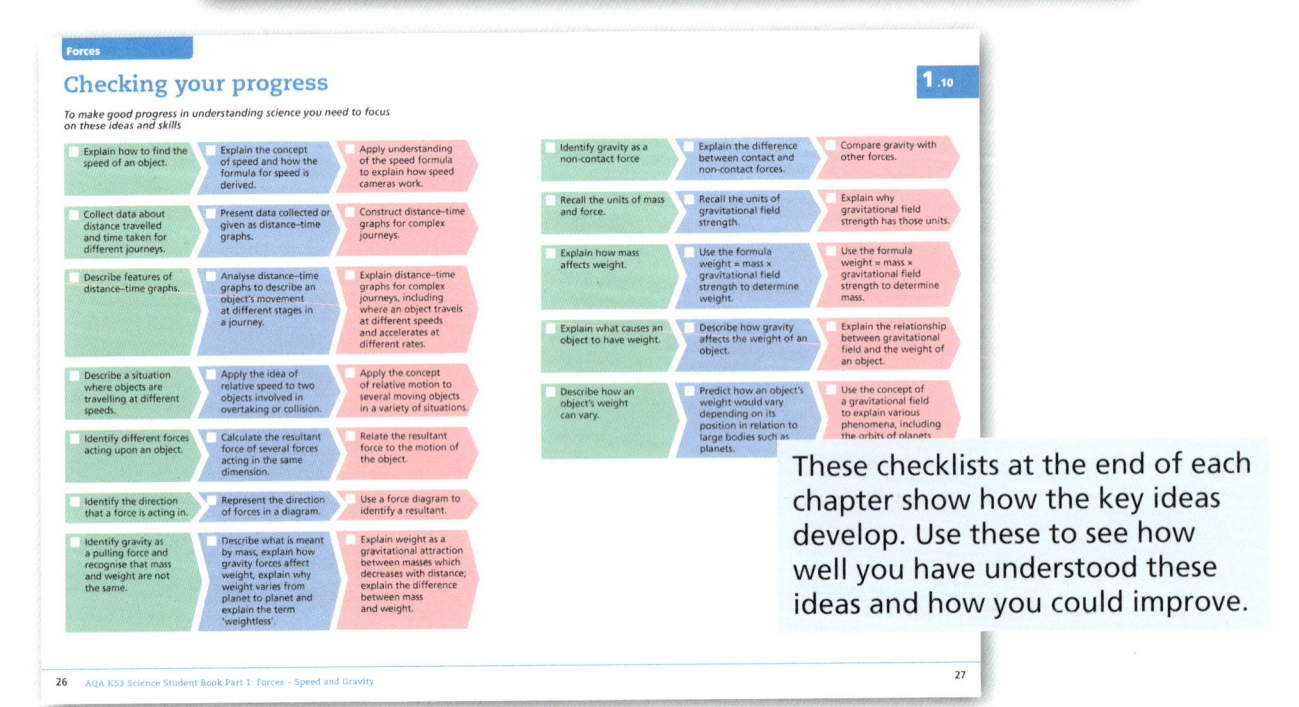

These checklists at the end of each chapter show how the key ideas develop. Use these to see how well you have understood these ideas and how you could improve.

AQA KS3 Science Student Book Part 1: How to use this book

This paragraph introduces the topic.

This tells you what you what you will learn.

Try the questions to check your understanding.

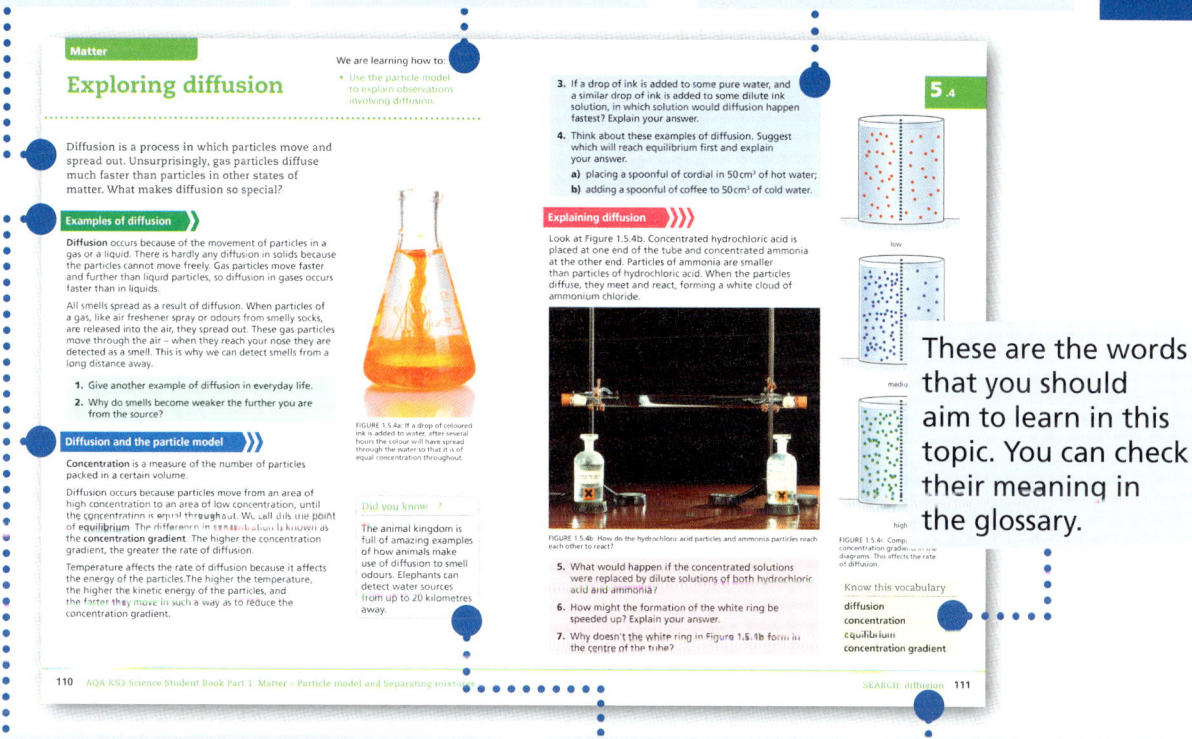

These are the words that you should aim to learn in this topic. You can check their meaning in the glossary.

Each topic is divided into three parts. The first section covers what you need to know, the second section helps you to apply your knowledge and the last section extends your knowledge and understanding.

These boxes contain interesting facts about the topic you are learning.

Key this phrase into an internet search box to find out more.

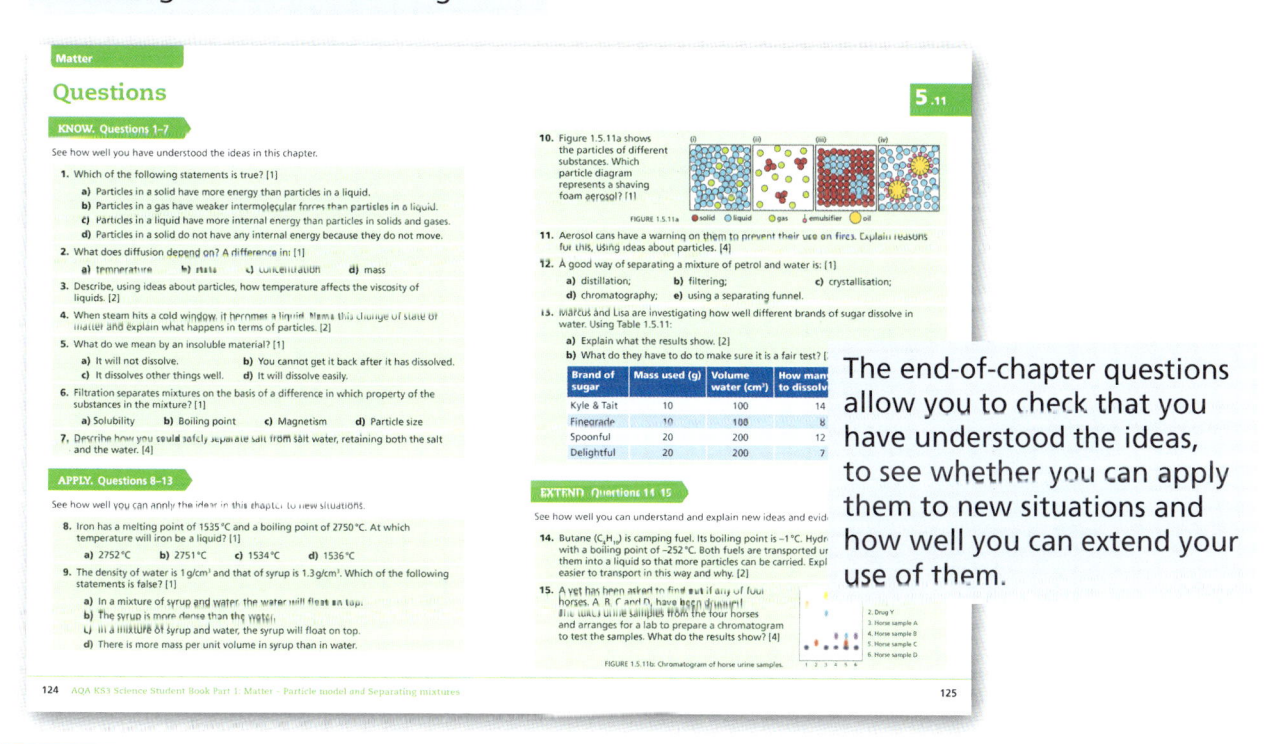

The end-of-chapter questions allow you to check that you have understood the ideas, to see whether you can apply them to new situations and how well you can extend your use of them.

5

Forces
Speed *and* Gravity

Ideas you have met before

Movement

Speed is a measurement of how quickly distance is being covered.

The speed of an object can be calculated by dividing the distance travelled by the time taken.

Speed is measured in units such as metres per second (m/s) and kilometres per hour (km/h).

Force

Forces can be pushes, pulls or turning forces. They can be 'contact' forces – when objects are touching – or 'non-contact' forces – when the forces act at a distance.

Force arrows drawn to scale show the size and direction of forces.

A newton-meter allows us to measure the size of a force.

Force is measured in newtons.

Gravity

Gravity is a non-contact force.

Large objects, like planets, exert strong gravitational forces on other objects. These objects are attracted towards the planet.

Gravity pulls objects towards the Earth.

Gravity keeps the Moon in orbit around the Earth and the Earth in orbit around the Sun.

Gravity affects objects such as people and rockets that are exploring space.

1.0

In this chapter you will find out »

Speed and acceleration

- The greater the speed, the shorter the time taken to cover a certain distance.
- An object's motion can be represented on a distance–time graph, which can be analysed to find out more about the motion.
- A straight line on a distance–time graph shows constant speed and a curved line shows acceleration.
- The motion of two objects can be compared and their relative speeds calculated.

Resultant force

- All the forces acting on an object can be combined to find the resultant – a single force which has the same effect.
- If the resultant force is not zero, the object will speed up, slow down or change direction.

Gravity

- Mass and weight are different, but related.
- Gravity is a non-contact force that acts between all masses.
- Every object exerts a gravitational pull on every other object.
- A planet, like the Earth, has a gravitational field.
- The gravitational fields of the Earth and other objects in the solar system affect space travel.

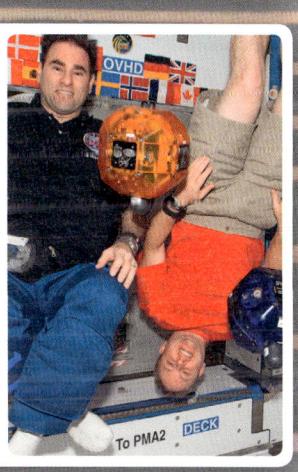

Forces

Understanding speed

We are learning how to:
- List the factors involved in defining speed.
- Describe a simple method to measure speed.
- Use the speed formula.

On Britain's busy roads, there are speed limits to make them safer. Driving too fast is one of the factors that causes accidents. Cameras that measure the speed of vehicles were introduced in the 1960s. In 2013 the number of deaths on Britain's roads was the lowest it had been since records began.

Distance and speed

When you travel on a journey, it takes a certain amount of time to travel the **distance**. The **speed** of a vehicle is worked out from how far a journey is and how long it takes. The **units** used for measuring speed are metres per second (m/s).

When travelling fast your speed is high. You cover a longer distance in a certain time – you travel more metres in each second, compared with travelling slower.

FIGURE 1.1.1a: Safety cameras and speed limit signs help to keep the number of deaths on British roads to a low level.

1. What does speed measure?
2. Which two quantities are needed to work out the speed at which something is travelling?
3. If car A travels 2 metres in one second and car B travels 2.5 metres in two seconds, which has the higher speed?
4. Motorbikes C and D both travel 100 metres. C takes 4 seconds and D takes 5. Calculate the speed of each motorbike.

FIGURE 1.1.1b: A car's speedometer shows the car's speed at each instant.

Calculating speed

1.1

We use a **formula** to calculate speed:

$$\text{speed} = \frac{\text{distance travelled}}{\text{time taken}}$$

The units of speed depend on which units were used for measuring the distance and the time.

Example calculation:
Usain Bolt from Jamaica won the 2016 Olympic 100-metre final in a time of 9.81 seconds.

speed = distance travelled ÷ time taken

Usain's speed = 100 ÷ 9.81 = 10.19 m/s
This is equivalent to about 37 km/h or about 23 mph.

> 5. Use the speed formula to calculate the speed of a cross-country runner who runs steadily for an hour and a half and covers 15 km. Show your working.
> 6. A mouse runs 2 metres in 4 seconds. What is its speed?

Did you know...?

A formula is a way of showing the relationship between quantities, using words or symbols.

Did you know...?

Some scientists have measured the force that an athlete's legs can produce, and how quickly the force can be transferred. From this they have worked out that it might be physically possible for the best athletes to run at over 60 km/h. We do not know if this will ever be achieved.

Average speed

When Usain Bolt won the Olympics sprint in 2016, his speed varied during the race. At the start it took a while to get up to full speed. The speed of 10.19 m/s that we calculated is his **average speed** over 100 metres. His top speed was over 12 m/s.

Some speed cameras work out a car's average speed over a distance of a kilometre or so, while other types work out speed almost in an instant. A car's speedometer displays the exact speed at any moment.

> 7. Explain why your average speed and your top speed over a car journey will be different.
> 8. What benefit to road safety may there be when cameras work out average speed over a distance, rather than in one spot?

FIGURE 1.1.1c: For an Olympic sprinter the distance is measured in metres (m) and the time is measured in seconds (s), so the speed is calculated in metres per second (m/s).

Know this vocabulary

distance
speed
unit
formula
average speed

SEARCH: measuring and calculating speed   9

Forces

Describing journeys with distance–time graphs

We are learning how to:
- Gather relevant data to describe a journey.
- Use the conventions of a distance–time graph.
- Display the data on a distance–time graph.

Science provides explanations for how the world works and gathers data to test the explanations. Graphs are a useful way of displaying data and can help you to understand the story behind the data.

Looking at distance–time graphs

The cyclists in Figure 1.1.2a are travelling at a steady speed along the path. This means that they cover the same distance every second.

The cyclists' journey can be represented on a **distance–time graph**, as shown in Figure 1.1.2b. For every second that passes, the cyclists travel 5 m. After 10 s they are 50 m from the starting point.

FIGURE 1.1.2a: Travelling at a steady speed.

You can use information from the graph to find how much distance has been covered at different times, how long it takes to travel different distances, and the cyclists' speed.

1. What unit should be used to measure the cyclists' speed in Figure 1.1.2b?
2. How far did the cyclists travel in the first 6 seconds of their journey?
3. What was the cyclists' speed?
4. Describe or sketch a line graph to show another cyclist who is travelling at half the speed. How does it differ from Figure 1.1.2b?

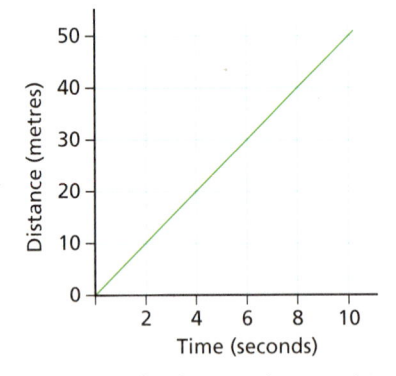

FIGURE 1.1.2b: Distance–time graph for constant speed.

Changing speed

In the distance–time graph in Figure 1.1.2c, the cyclist does not travel the same distance every second. For the first 10 s they travel at a slow speed and cover little distance. However, they gradually **accelerate** (speed up).

The speed between 30s and 45s is faster than before because the cyclist covers more distance every second. The steeper line of the graph indicates that the speed has increased. Subsequently the cyclist stops and then remains **stationary**. The flat part of the graph shows that no more distance is covered.

5. On a distance–time graph, what does it mean when:
   a) the graph is a horizontal line?
   b) the graph is a straight upward-sloping line?
   c) the graph is an upward-sloping curve?

6. Looking at Figure 1.1.2c, how long was the cyclist stationary for?

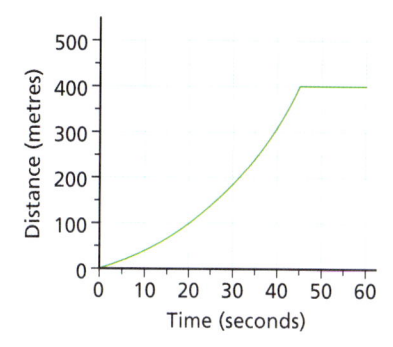

FIGURE 1.1.2c: A distance–time graph for a cyclist who changes speed.

Complex journeys

Figure 1.1.2d shows a distance–time graph for a student's journey to school which includes walking (1), waiting for a friend (2), walking with their friend (3), waiting for a bus (4) and riding on the bus (5). Their speed varies during different sections of the journey – at certain times no distance is covered.

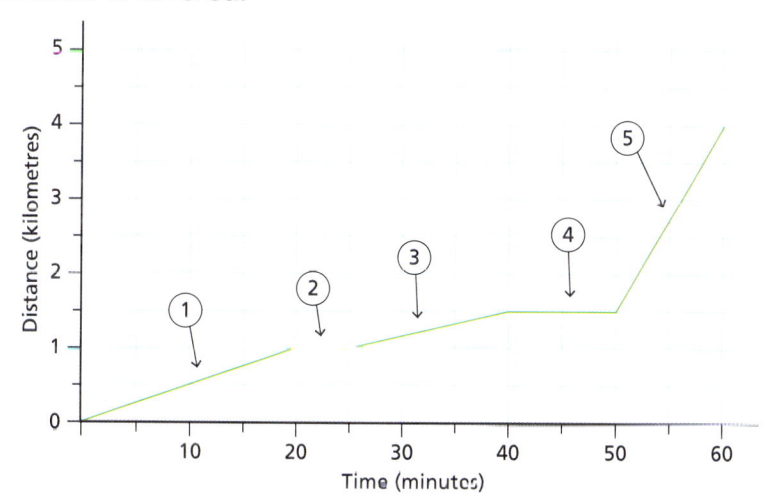

FIGURE 1.1.2d: A distance–time graph for a student's journey to school.

7. Looking at Figure 1.1.2d, what is the evidence that the students travelled faster on the bus than at other times during the journey?

8. Compare sections 1 and 3 on the graph. How are they different? Suggest a possible reason for the difference.

9. Imagine a journey where you travel from your home to an overseas holiday destination. Sketch a line graph to represent the journey. Label each part of the graph to explain what is happening.

Did you know...?

From a distance–time graph you can work out the average speed for a whole journey during which the speed varies at different times. You can also work out the speed at different parts of the journey.

Know this vocabulary

distance–time graph
accelerate
stationary

SEARCH: distance–time graphs

Forces

Exploring journeys on distance–time graphs

We are learning how to:
- Interpret distance–time graphs to learn about the journeys represented.
- Relate distance–time graphs to different situations and describe what they show.

A speeding motorist sees a speed camera and slows down. The car then accelerates and is again breaking the speed limit. Further along the road a second camera comes into view and, again, the driver slows. A few days later a letter from the police arrives in the post…

Speed cameras and distance–time graphs

Speed measurement on roads is often done by cameras that record the position of a car at the start and at the end of a period of time. The further the distance the car moved during that time, the faster it was going. It is then simple to use the speed formula to calculate the car's speed.

FIGURE 1.1.3a: A speed camera.

Motorists who realise they are speeding may suddenly slow down when they see a camera. Figure 1.1.3b shows what a distance–time graph might look like in such a situation.

1. What is the formula for calculating speed?
2. Looking at Figure 1.1.3b, how can you tell that the car's speed has changed?
3. Calculate the speed of the car in both sections of the graph. Show your working.

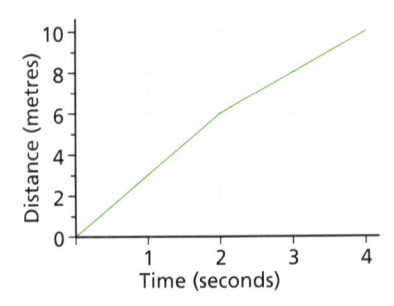

FIGURE 1.1.3b: A distance–time graph for a car approaching a speed camera.

Using a time-lapse sequence

Figure 1.1.3c shows a **time-lapse sequence** taken as a candle burned. The photographs were taken at five-minute intervals. The candle burned at a steady rate so it got shorter by a similar amount every five minutes.

The same process of time-lapse photography can be used to record the motion of objects, such as cars. The longer the distance between a car's position in successive photographs, the faster it must have been travelling. For example, if photographs are taken at one-second intervals and a car moves 12 m between each photograph, then the speed of the car is 12 m/s.

Did you know…?

A time-lapse sequence is a series of images showing the same object or scene at different points in time.

FIGURE 1.1.3c: Time-lapse photography of a burning candle – images were taken every 5 minutes.

1.3

4. Look at Figure 1.1.3c. How long did the candle take to burn?
5. Read the final paragraph on the opposite page. What is the speed of the car in km/h?
6. Assuming the car in question 5 is travelling at a steady speed, draw a distance–time graph to show its motion.

Acceleration

If you consider all the forces acting on an object and they don't cancel each other out then the object will accelerate (increase in speed). Later on in this chapter we'll be finding out more about force and acceleration.

On a distance–time graph, a steep slope shows that an object is travelling faster than an object with a shallow slope. If the object rapidly accelerates, the slope of the line will change rapidly. However, if the change in speed is more gradual, the gradient will change more gradually.

Figure 1.1.3d shows three different journeys. The change in each slope shows how quickly the speed of each object changes. How quickly speed changes is called **acceleration**.

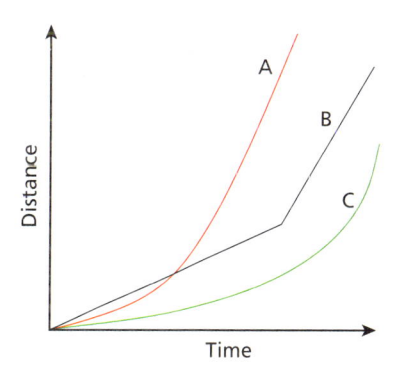

FIGURE 1.1.3d: Looking at acceleration on a distance–time graph.

7. Looking at Figure 1.1.3d, which object has:
   a) the fastest speed at the start?
   b) the fastest final speed?
8. If speed is measured in m/s, what unit is used for acceleration?
9. Sketch the graph in Figure 1.1.3d and extend the lines to show:
   a) object A continuing at the same speed for a while and then stopping abruptly;
   b) object B coming to a gradual halt having travelled a shorter total distance than object A;
   c) object C slowing down and then travelling at a steady speed.

Did you know…?

Drag race cars can accelerate from a standstill to cover a 300 m straight-line race track in less than 4 s, reaching speeds of over 500 km/h.

Know this vocabulary

time-lapse sequence
acceleration

SEARCH: acceleration

Forces

Investigating the motion of a car on a ramp

We are learning how to:
- Describe the motion of an object whose speed is changing.
- Devise questions that can be explored scientifically.
- Present data so that it can be analysed to answer questions.

A toy car released at the top of a ramp will accelerate. We can explore what will make it go faster and whether a steeper ramp necessarily means a longer journey.

Thinking about the journey

If we put a toy car at the top of a ramp and release it, it will accelerate down the ramp. When it reaches the end of the ramp it will continue moving but slow down. The acceleration at the start of the journey is caused by forces acting on the car. It eventually stops because of forces too.

1. Explain why the car accelerates down the ramp.
2. Explain why the car decelerates once it reaches the ground.
3. Suggest why the forces in the first part of the journey make it accelerate and those in the second part make it decelerate.

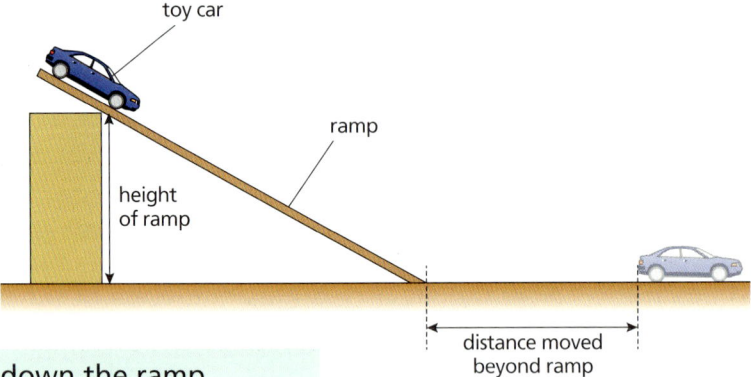

FIGURE 1.1.4a: The car's journey.

Planning the investigation

When we plan an investigation like this, we need to think about the things we could alter that might affect the motion. It will be important to identify these because these are factors we might want to investigate. We might want to find out what happens if we change them. Alternatively, we might want to identify them so that we can keep them the same. If the motion changes, we will want to know what we've done to cause that. If we've changed several things, we won't know which of them has made a difference.

14 AQA KS3 Science Student Book Part 1: Forces – Speed *and* Gravity

1.4

Things we could change are called **independent variables**. Altering these could make a difference to other things, which are called **dependent variables**. Some of the independent variables will be kept the same; these are then called **control variables**. If we alter one independent variable and see how a dependent variable changes we can look for a **correlation**.

4. a) If you are setting up a toy car to roll down a ramp, what are the independent variables?

   b) What might change as a result? (These are the dependent variables.)

5. a) Select one independent variable to alter and a dependent variable to measure. Use these to write an enquiry question in the form 'How does ... affect ...?'

   b) In this case, what will the control variables be?

Presenting data

Altering the independent variable means selecting values. If you changed the height of the ramp, for example, you need to decide which heights to use. When you have these values you can set up a table showing this data and also the dependent variable, which changed as a result.

Some data is continuous, because it can take any value. Changing the mass of the car by adding modelling clay gives continuous data, because any amount can be added. Other data is discrete because it can only have certain values. Swapping the red car for a blue car gives discrete data. From your table of data you can draw a graph. You will need to decide what kind of graph is appropriate.

If the independent variable is discrete, you should use a bar chart. If it is continuous, use a line graph.

6. What headings would your table need?

7. Would you need to repeat any of the readings?

8. What conclusion could you draw from your graph?

9. Ali and Dave decided to explore the relationship between the height of the ramp and the distance travelled by the car after it reached the end of the ramp. Sketch the graph they might have got.

Know this vocabulary

independent variable
dependent variable
control variable
correlation

SEARCH: investigating speed

Forces

Understanding relative motion

We are learning how to:
- Describe the motion of objects in relation to each other.
- Explain the concept of relative motion.
- Apply the concept of relative motion to various situations.

Imagine driving along a motorway. Alongside your car is another car travelling at exactly the same speed. Both cars' speedometers could be reading over 100 km/h, but compared to each other the cars are not moving at all.

Relative motion

When scientists compare the movement of two objects, they talk about **relative motion**. For example, if a car is travelling at 50 km/h and is being caught by a car doing 55 km/h, the speed of the second car relative to the first – its **relative speed** – is 5 km/h.

If you compare a cyclist doing 20 km/h and a car doing 60 km/h, the car is travelling at 40 km/h relative to the cyclist. After 1 hour the car has travelled 40 km further than the cyclist.

FIGURE 1.1.5a: The car travels faster in relation to the bicycle.

1. A person sets off jogging along a canal path at 12 km/h at the same time as a boat sets off at 10 km/h.
   a) How far will each one travel in half an hour?
   b) What is their relative speed?
   c) To the jogger, how would the boat appear to be moving as they travel along the canal?

Journeys and collisions

Figure 1.1.5b shows the distance–time graphs for two cars on a motorway. Car B set off later than car A. You can see when each will have completed their journey and the distance between them.

If the cars were in the same lane, car B would crash into the back of car A. It is the relative speed of two cars in a collision that is important rather than the actual speed of one car alone.

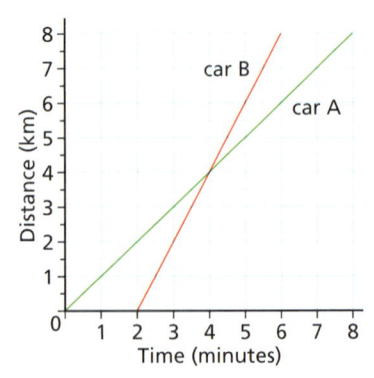

FIGURE 1.1.5b: Two cars travelling at different speeds.

1.5

Two cars travelling at 40 km/h towards each other have a relative speed of 80 km/h. This is equivalent to a moving car approaching a stationary car at 80 km/h.

2. Look at Figure 1.1.5b. What are the speeds of the two cars in km/h? What is their relative speed?
3. What is the relative position of the two cars:
    a) 2 minutes after car A sets off?
    b) 1 minute later?
4. Explain why head-on collisions are so dangerous.

Looking at events differently »»»

If you look at the sky from a moving car it can be very difficult to tell which way the clouds are moving. They can appear to be stationary if the car is travelling at the same speed as the clouds. If the car speeds up, the clouds may appear to the passengers to be travelling in the opposite direction to the car.

FIGURE 1.1.5c: Two cars travelling at different speeds.

FIGURE 1.1.5d: The relative motion depends on the speed and direction of the car and clouds.

5. Explain why in some situations it is hard to tell whether or not you are moving. How could your other senses help your judgement?
6. Explain the similarities and differences between these situations:
    a) a car travelling at 10 km/h and colliding with a parked car;
    b) a car travelling at 70 km/h and colliding with a car doing 60 km/h in the same direction;
    c) a car travelling at 70 km/h and colliding with a car doing 60 km/h in the opposite direction.

Did you know…?

If you travel away from a loud noise faster than 344 m/s, you will never hear the sound. Sound travels through air at just over 343 m/s, so it would never catch you up.

Know this vocabulary

relative motion
relative speed

SEARCH: relative motion

Forces

Understanding forces

We are learning how to:
- Recognise different examples of forces.
- List the main types of force.
- Represent forces using arrows.

All forces try to pull, push, twist or break objects but some are contact forces and some are non-contact forces. Gravity, like magnetism, is a non-contact force. It can have a spectacular effect though.

Types of force

A **force** can be a pushing force, a pulling force or a turning force. There is a pulling force from the Earth on this bungee jumper. Once he steps off the platform, the pulling force makes him fall. The arrow shows the pulling force making him move downwards. Without the pulling force of the Earth, he would not fall. The pulling force of the Earth on objects is called **gravity**.

1. How would you describe the type of force that the Earth produces on the bungee jumper?
2. What is the name given to this force?

FIGURE 1.1.6a: The downward force acting on a bungee jumper.

Multiple forces

A number of forces can be acting on something at the same time. The aeroplane in Figure 1.1.6b has four main forces acting on it:

- the downward pull of gravity;
- the forward push from the engines;
- the upward pull provided by the lift from the wings;
- the pushing force of the air which resists the plane as it moves.

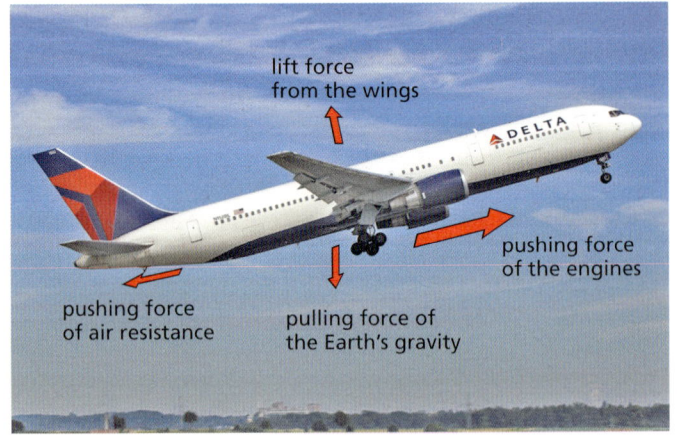

FIGURE 1.1.6b: These forces act on an aeroplane as it takes off.

The direction of a force can be shown by an arrow. We can show how strong one force is compared to another by using different-sized arrows.

18 AQA KS3 Science Student Book Part 1: Forces – Speed *and* Gravity

3. Which forces are helping the plane in Figure 1.1.6b to fly?

4. Which forces are working against the plane when it flies?

5. Draw and label a force diagram showing the aircraft in horizontal flight.

Forces in balance

The two tug-of-war teams in Figure 1.1.6c are pulling equally and no one is moving. All the forces are in **balance**, which means each force is perfectly balanced by an equal force in the opposite direction. If we combine all the forces acting in a situation we can find the resultant force. Think of this as a single force which could replace all the others and have the same effect. In this situation the resultant force is zero.

FIGURE 1.1.6c: Forces are present, but there is no movement.

6. What would happen to the size and direction of the resultant force if an extra person were added to the left-hand team in Fig 1.1.6c?

7. Sketch a car that is starting to move away from a set of traffic lights. Draw arrows to show the forces at work and comment on the direction of the resultant force.

8. Draw force diagrams and calculate the size and direction of the resultant force if:

    a) a boat has a force of 500 N from the wind pushing it forwards and the water resistance is 200 N;

    b) a sledge is being pulled with a force of 250 N and acted on by friction (100 N) and air resistance (50 N).

Did you know…?

When forces are balanced their size and direction cancel each other out.

Did you know…?

When you see films of astronauts inside a space station **orbiting** the Earth, the astronauts appear to be weightless. But they, and the space station and everything in it, are actually still being attracted by the Earth's gravity. If there were no pulling force of gravity from the Earth, the space station would fly off into space.

Know this vocabulary

force
gravity
balance
orbit

SEARCH: types of forces

Forces

Understanding gravitational fields

We are learning how to:
- Describe gravity as a non-contact force.
- Explore the concepts of gravitational field and weight.
- Explain how weight is related to mass.

The Earth's gravitational field

The region around the Earth affected by its gravity is its **gravitational field**. A **field** is an area in which an object feels a force.

Within the Earth's gravitational field objects are pulled towards the Earth. This pull is a **non-contact force** because it acts at a distance – objects do not have to be on a planet's surface to be affected.

1. In what direction does Earth's gravitational force act?
2. Describe what is meant by a gravitational field.

Gravitational field strength and weight

The Earth's **gravitational field strength** gets weaker the further you move from the Earth's surface. It also varies slightly in strength across the surface. Other planets and moons have similar gravitational fields.

Gravitational fields can extend over long distances. Even though the Moon is over 350 000 km from Earth, they are affected by each other's gravitational fields.

Gravity does not stop at the Earth's surface. If you descend into a deep mine you are still pulled towards the middle of the Earth.

The **weight** of an object depends on the mass of the object and the strength of the gravitational field acting on it. The formula used to calculate weight is:

weight of object (W) = mass (m) of object × gravitational field strength (g)

Weight is measured in newtons (N) and mass is measured in kilograms (kg), so the gravitational field strength is measured in newtons per kilogram (N/kg).

On the surface of the Earth the gravitational field strength (symbol g) is about 10 N/kg. To calculate how much a bag of fruit with a mass of 2 kg would weigh on the Earth's surface:

$W = m \times g = 2 \times 10 = 20$ N

FIGURE 1.1.7a: Gravity acts all over the Earth towards its centre.

FIGURE 1.1.7b: The rise and fall of the tide is largely due to the Moon's gravitational field.

AQA KS3 Science Student Book Part 1: Forces – Speed *and* Gravity

3. What evidence exists that the Moon's gravitational field affects the Earth?
4. List the main differences between the pulling force due to gravity and the pulling force from a rope in a tug-of-war.
5. Which quantities determine the weight of an object?
6. Calculate, if g = 10 N/kg:
    a) the weight of a 25 kg mass;
    b) the mass of a 1000 N weight.
7. Explain why the weight of an object can vary, but the mass always stays the same.

Acceleration in gravitational fields

The pulling force on an object in a gravitational field causes it to accelerate in the direction of the force. The stronger the field, the bigger the acceleration – they have the same numerical value. For example, on the Earth's surface the field strength of 10 N/kg causes an unsupported object to accelerate towards the Earth at 10 m/s^2. The acceleration depends on the gravitational field strength but not on weight or mass.

When investigating acceleration in the Earth's gravitational field, other factors such as air resistance can affect the results.

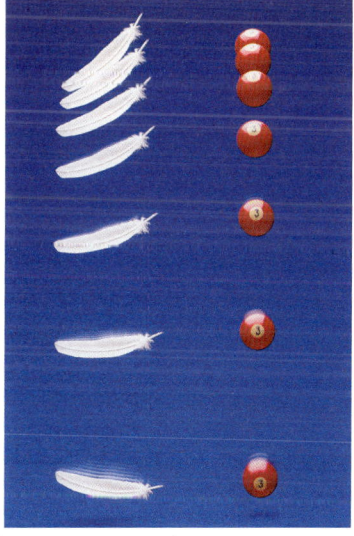

FIGURE 1.1.7d: Do different masses really fall at the same rate?

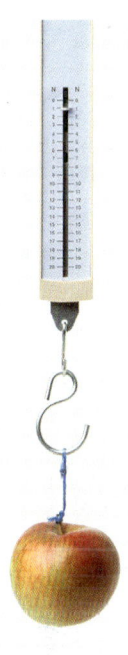

FIGURE 1.1.7c: All masses close to the Earth's surface are pulled by the gravitational field strength of 10 N/kg.

Did you know…?

The 'rule' about all objects falling at the same rate applies when there is no air resistance. Air resistance has a different effect on objects of different mass and different shape.

8. Draw force diagrams showing:
    a) an apple suspended by a balance;
    b) an apple in free fall.
9. Different masses fall towards the Earth at the same rate if air resistance is not a factor – explain why.
10. Design an activity to find out if air resistance affects the rate at which objects fall.

Know this vocabulary

gravitational field
field
non-contact force
gravitational field strength
weight

SEARCH: gravitational fields

Forces

Understanding mass and weight

We are learning how to:
- Explain the difference between mass and weight.
- Apply ideas of weight to space travel.

We all know about weight because we all have it. It's caused by gravity, so we wouldn't have as much weight if we went to the Moon. We need to understand about mass too. It's linked with weight but isn't so easy to lose.

Weight, gravity and mass

The **weight** of an object is the force of **gravity** pulling down on the object. If there were no gravity then everything would be weightless. Because weight is a force, it should be measured in newtons. We can measure the weight of an object using instruments such as newton-meters and bathroom scales. Both measure how much the object is pulled down by gravity.

FIGURE 1.1.8a: Someone 'weighing' themselves with bathroom scales.

Mass is a measure of the amount of material in an object – the number of particles and type of particles it is composed of. Mass does not depend on the force of gravity, so it does not change if you take it somewhere where the gravitational field is not as strong, such as the Moon. Mass is measured in kilograms. The mass of an object can be measured using a balance that compares the object with a known mass.

FIGURE 1.1.8b: Using a balance to find the mass of an object.

Sometimes people mix up 'mass' and 'weight', so scientists need to be careful to choose which term to use.

1. Why do you think that some people confuse weight and mass?
2. If you measured the mass and the weight of an object on two planets of different sizes, what differences would you notice? Explain your answer.

Gravity in space

1.8

The force of gravity on you (your weight) depends on your distance from a planet. The further away you are from the Earth, the weaker the gravitational field strength, so the weaker the force pulling you back. In outer space, the distance to the nearest planets and stars could be so big that there would be no noticeable force of gravity – you would be weightless.

> 3. In much of outer space there is little or no gravity. Why is this?
>
> 4. Think of a spacecraft setting off from Earth and travelling directly to the Moon. Describe the changes in gravity you expect the spacecraft to experience during the journey.

Gravity in orbit

In videos of the astronauts in the International Space Station, in **orbit** around the Earth, the astronauts look as if they have no weight. However, the Earth's gravitational field is pulling on them and also on the space station and everything in it. They fall at the same rate so inside the station the astronauts float about. They appear to be – and feel – weightless.

Did you know…?

An object in orbit, such as a space station, is not in zero gravity. It is still being attracted by the Earth's gravity. If there were no pulling force of gravity from the Earth, the space station would fly off into space.

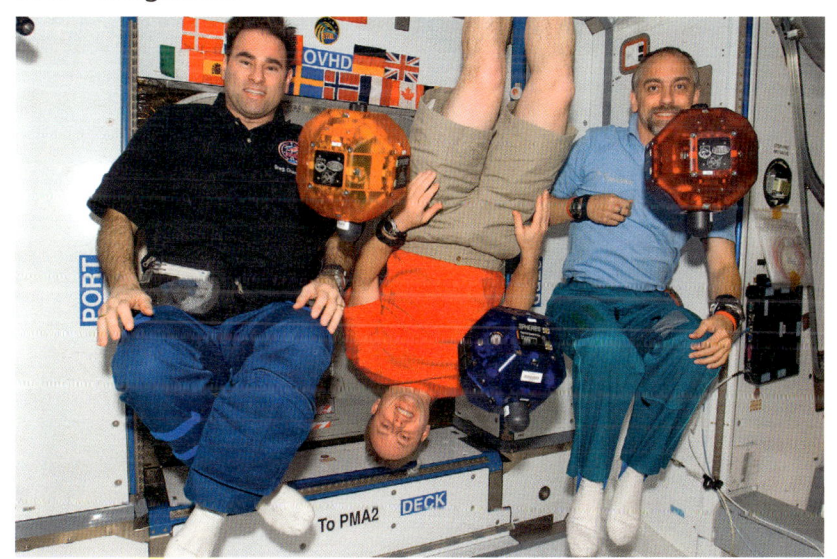

FIGURE 1.1.8c: These astronauts in the space station are falling at the same rate as the space station.

> 5. Compared to standing on Earth, what would your weight be on a high-flying plane?
>   a) Stronger b) The same c) Weaker d) Zero
>
> 6. Explain your answer to question 5.

Know this vocabulary

weight
gravity
mass
orbit

SEARCH: the difference between mass and weight

Forces

Understanding gravity

We are learning how to:
- Understand that gravity varies according to where you are in the solar system.
- Apply ideas about gravity to various situations.

We are all experts in coping with gravity. We've dealt with it all our lives. We've made use of it as children on swings and slides. We know the problems it causes when we drop glass or china on a hard floor. But what is it?

Gravity and space exploration

When people explore space, one of the problems they face is coping with the way **gravity** varies. Astronauts visiting the Moon weighed much less than on Earth and in deep space they would be weightless. **Weight** depends on the force of gravity from massive objects such as stars, planets and moons.

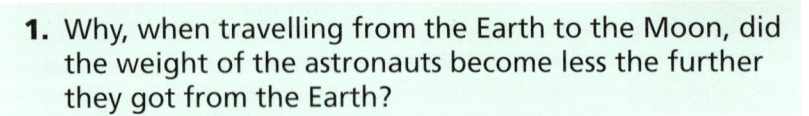

FIGURE 1.1.9a: The Earth and the Moon. The bigger the **mass** of a planet or moon, the stronger its force of gravity.

1. Why, when travelling from the Earth to the Moon, did the weight of the astronauts become less the further they got from the Earth?
2. How did their weight change when they got nearer to the Moon?
3. Why do you think they weighed less on the Moon?

Understanding gravity

Gravity is a force that pulls pairs of objects together. For example, your body is pulled towards the Earth, and the Earth and other planets are held in orbit around the Sun.

Gravity actually exists between *all* objects, but the force is only large enough to be noticeable when a massive object, such as a planet or a star, is involved.

4. Look at Table 1.1.9. Where is gravity greatest?
5. Using information from the table, write the planets and the Moon in order of increasing gravitational field strength, if you were standing on the surface.

Did you know…?

Trying to stay fit is a real challenge in a location where gravity is less. If your muscles don't have to work to support you, you soon become unfit. The International Space Station has exercise machines which astronauts use regularly.

1.9

6. The table shows how high you could jump, in each of these places, using the same amount of force. Use the idea of opposing forces to explain why it varies so much.

TABLE 1.1.9: The effects of different values of gravity on the Moon and on other planets in the solar system.

	Earth	Moon	Mercury	Venus	Mars
Surface gravity (compared with the Earth's)	1.00	0.17	0.38	0.90	0.38
Your mass (compared with your mass on Earth)	1	1	1	1	1
How much you can lift (kg)	10	60	30	10	30
How high you can jump (cm)	20	120	53	22	53
How long it takes to fall back to the ground (s)	0.4	2.4	1.1	0.4	1.1

A gravity puzzle

Gravity is an attractive (pulling) force between masses. What gravity would you experience if you tunnelled towards the centre of the Earth?

Under the surface there would be a force of gravity from the mass of the Earth above you as well as from that below you. Because these forces are in opposite directions, the overall force of gravity would be lower than on the Earth's surface.

7. Imagine it was possible to build a tower on Earth to the height of an orbiting space station.
    a) What force(s) would you experience if you stepped off the tower?
    b) What movement would you expect?
8. Explain what would happen if you tried to weigh yourself in these situations:
    a) outer space;
    b) in a tunnel, halfway to the Earth's centre;
    c) on top of a tower at space station level.

Know this vocabulary

gravity
weight
mass

SEARCH: gravity in the solar system

Forces

Checking your progress

To make good progress in understanding science you need to focus on these ideas and skills

☐ Explain how to find the speed of an object.	☐ Explain the concept of speed and how the formula for speed is derived.	☐ Apply understanding of the speed formula to explain how speed cameras work.
☐ Collect data about distance travelled and time taken for different journeys.	☐ Present data collected or given as distance–time graphs.	☐ Construct distance–time graphs for complex journeys.
☐ Describe features of distance–time graphs.	☐ Analyse distance–time graphs to describe an object's movement at different stages in a journey.	☐ Explain distance–time graphs for complex journeys, including where an object travels at different speeds and accelerates at different rates.
☐ Describe a situation where objects are travelling at different speeds.	☐ Apply the idea of relative speed to two objects involved in overtaking or collision.	☐ Apply the concept of relative motion to several moving objects in a variety of situations.
☐ Identify different forces acting upon an object.	☐ Calculate the resultant force of several forces acting in the same dimension.	☐ Relate the resultant force to the motion of the object.
☐ Identify the direction that a force is acting in.	☐ Represent the direction of forces in a diagram.	☐ Use a force diagram to identify a resultant.
☐ Identify gravity as a pulling force and recognise that mass and weight are not the same.	☐ Describe what is meant by mass, explain how gravity forces affect weight, explain why weight varies from planet to planet and explain the term 'weightless'.	☐ Explain weight as a gravitational attraction between masses which decreases with distance; explain the difference between mass and weight.

1.10

- ☐ Identify gravity as a non-contact force
- ☐ Explain the difference between contact and non-contact forces.
- ☐ Compare gravity with other forces.

- ☐ Recall the units of mass and force.
- ☐ Recall the units of gravitational field strength.
- ☐ Explain why gravitational field strength has those units.

- ☐ Explain how mass affects weight.
- ☐ Use the formula weight = mass × gravitational field strength to determine weight.
- ☐ Use the formula weight = mass × gravitational field strength to determine mass.

- ☐ Explain what causes an object to have weight.
- ☐ Describe how gravity affects the weight of an object.
- ☐ Explain the relationship between gravitational field and the weight of an object.

- ☐ Describe how an object's weight can vary.
- ☐ Predict how an object's weight would vary depending on its position in relation to large bodies such as planets.
- ☐ Use the concept of a gravitational field to explain various phenomena, including the orbits of planets around stars.

Forces

Questions

KNOW. Questions 1–5

See how well you have understood the ideas in this chapter.

1. What is the speed of a cyclist who covers 7 m in 1 second? [1]
   a) 70 km/h   b) 7 km/h   c) 7 m/s   d) 700 m/s

2. Which one of these units is *not* used for speed? [1]
   a) km/h   b) m/s   c) N/kg   d) mph

3. Which row of the table shows the correct units? [1]

   TABLE 1.1.11

	Mass is measured in ...	Weight is measured in ...
A	N	N
B	N	kg
C	kg	N
D	kg	kg

4. Which of these is true about gravity on the Moon? [1]
   a) There is no gravity on the Moon.
   b) There is gravity on the Moon but it's less than on the Earth.
   c) Gravity is the same strength everywhere – it's universal.
   d) Gravity is greater on the Moon than on the Earth.

5. Which of these statements about weight is *not* true? [1]
   a) Weight is affected by the gravitational field strength in the area.
   b) Weight is affected by the mass of an object.
   c) Weight increases at greater distances from massive objects.
   d) Weight is affected by where an object is.

APPLY. Questions 6–10

See how well you can apply the ideas in this chapter to new situations.

6. A van takes 1 hour to travel along a 60 km stretch of road. A car takes 45 minutes to do the same journey. Which of these statements is true about their relative speeds? [1]
   a) The van is 15 km/h faster.
   b) The van is 15 km/h slower.
   c) The car is 20 km/h faster.
   d) The car is 20 km/h slower.

7. Sketch a distance–time graph for two horses moving across a 500 m field. One horse trots steadily across the field in 4 minutes. The other accelerates to a gallop, stops for 1 minute to eat grass and then gallops the rest of the way, reaching the far side after 3 minutes. [4]

8. Which of these statements is true about your weight and mass on a planet that has twice the gravitational field strength of Earth? [1]
   a) Weight is the same, mass is double.
   b) Weight and mass are both the same.
   c) Weight and mass are both double.
   d) Weight is double, mass is the same.

9. In which of these locations would the gravitational field be the strongest? [1]
   a) On the surface of a red giant star.
   b) On the Moon as it orbits the Earth's surface.
   c) At the edge of the Earth's atmosphere.
   d) On the Earth's surface.

10. Which of these statements would be true for a rocket taking off from a launch pad and accelerating towards space? [1]
    a) The forces on it are balanced.
    b) There will be a resultant force acting upwards.
    c) Whilst it's travelling vertically upwards it will not experience air resistance.
    d) Once it gets into orbit around the Earth, gravity will no longer act on it.

EXTEND. Questions 11–13

See how well you can understand and explain new ideas and evidence.

11. Which of these defines what the strength of the gravitational field on the surface of a planet depends on? [1]
    a) The mass and radius of the planet.
    b) The mass of the planet and its distance from the Sun.
    c) The shape and radius of the planet.
    d) The rotation speed of the planet.

12. Draw a diagram to show what would happen to a satellite if the Earth's gravity were suddenly turned off. [1]

13. Scientists believe that there is a massive black hole at the centre of our galaxy, with a mass 30 billion times that of the Sun. Explain what evidence could suggest its existence.

    If a spacecraft set off in a straight line at a constant speed towards the black hole, what additional evidence would indicate the presence of the black hole?

    What challenges would the spacecraft face? [4]

Electromagnets
Voltage and resistance *and* Current

Ideas you have met before

Components in a circuit

All metals are good electrical conductors. Materials that do not allow electricity to pass through them are called insulators. Examples are wood, plastic, rubber, cloth and air.

A simple electric circuit consists of components such as cells, wires, bulbs, switches and buzzers.

Recognised symbols can be used to represent a simple circuit in a diagram.

Making current flow

Components only work if the circuit is complete and contains a power supply. Then an electric current can flow.

When the switch is open (off), the circuit is not complete and none of the components will work.

Changing the voltage

The brightness of a lamp or the loudness of a buzzer is related to the number and voltage of cells used in the circuit.

If more cells are added to a circuit, the brightness of bulbs or the loudness of buzzers in the circuit will increase.

In this chapter you will find out

2.0

Explaining electric circuits

- Components in an electric circuit provide opposition to the current, known as resistance, and transfer energy to the surroundings.
- Components in circuits can be arranged in series or in parallel. These arrangements have different effects on the voltage and current, and provide different applications.
- The current, voltage and resistance are related to each other
- Models are a good way of explaining what happens in a circuit.

Current

- Current is a movement of electrons and is the same everywhere in a series circuit.
- Current depends on the 'push' given by the battery, known as the voltage.
- Current divides between loops in a parallel circuit and combines when loops meet.

Potential difference

- Voltage, or 'potential difference', is the amount of energy per unit of charge transferred through the electrical pathway.
- In a series circuit, voltage is shared between each component. In a parallel circuit, voltage is the same across each loop.

Electrostatic force

- Around a charged object, the electric field affects other charged objects, causing them to be attracted or repelled.
- The field strength decreases with distance.

Electromagnets

Describing electric circuits

We are learning how to:

- Describe circuits and draw circuit diagrams.
- Explain what is meant by current.
- Explain how materials allow current to flow.

A light bulb in an electric circuit lights up instantaneously. Even if the circuit were the size of a football pitch, there would be no time delay for the light to come on. What is actually going on in the circuit for energy to be transferred so quickly?

Components in electric circuits

An electric circuit is a loop of wire with its ends connected to an energy source, such as a battery or cell. Strictly, a 'battery' is two or more cells together.

When a circuit is complete, energy is transferred from the battery to the wires by a flow of charge that we call an electric current. Devices such as light bulbs, motors and buzzers are **components** that can make use of this energy transferred.

If there are any gaps in the circuit, the current will not flow and energy cannot be transferred. A material that allows current to pass through it is called an **electrical conductor**. These contain small charged particles called **electrons** that are free to move within the conductor. An **electrical insulator** does not have any free electrons and cannot allow a current to pass.

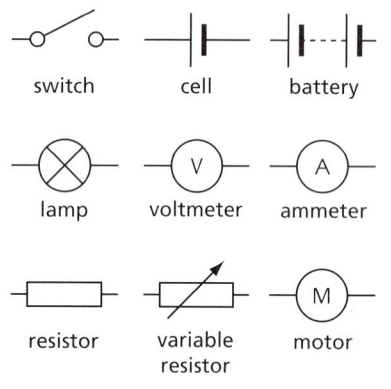

FIGURE 1.2.1a: Circuit symbols for common components.

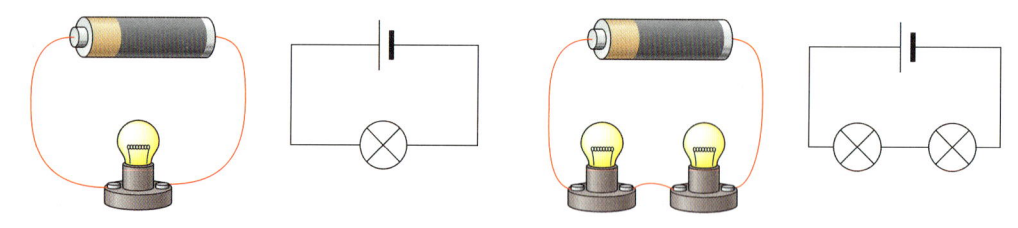

FIGURE 1.2.1b: How circuit symbols are used to represent components in a circuit diagram.

1. If pencil lead is placed in a circuit with a light bulb, the bulb lights up. What conclusion can you draw about this material?

2. Draw a circuit diagram for a circuit with one cell and three bulbs.

3. Why is it important to represent components with symbols?

Using models to explain current

2.1

Current is the rate of flow of charge (electrons) in the circuit, and is given the symbol *I*. It is measured by an **ammeter** in **amperes** (symbol A), after the French scientist André-Marie Ampère.

Models and analogies are often used to explain complex phenomena like current. One analogy is to compare electric current to water flowing in a stream. The charges are the water particles, and the current is the flowing stream.

Another analogy used to represent current is that of a convoy of coal trucks. The trucks represent the charged particles, the movement of the trucks represents the current, and the coal they carry represents the energy they transfer.

FIGURE 1.2.1c: In the analogies pictured in the photos, what represents the charge and what represents the current?

4. Using first the water analogy and then the coal-truck analogy, draw diagrams to show the difference between a low current and a high current.

5. Which analogy is better at explaining that current transfers energy to different components? Explain your answer.

Scientific explanation of current

When the battery is connected, the electrons in all parts of the wires within the circuit move at the same time, in the same direction and at the same rate. This movement constitutes the current. In this way, no matter where the components are in the circuit, they will all conduct at the same time – there is no delay because all the electrons in the circuit move simultaneously.

Current is not used up in the circuit. It has the same value before and after each component in the circuit.

6. Explain the strengths and limitations of the two analogies above, in light of the scientific explanation for current.

7. Explain why current is not used up in a circuit.

Did you know…?

A current of 1 amp means there are 6 250 000 000 000 000 000 electrons flowing past a point every second!

Know this vocabulary

component
electrical conductor
electrons
electrical insulator
current
ammeter
ampere

SEARCH: electric current   33

Electromagnets

Understanding energy in circuits

We are learning how to:
- Describe what voltage does in a circuit.
- Recall how voltage can be measured.
- Explain the effect of increasing the voltage supplied.

We know that an electric circuit gets its energy from a cell or battery. The amount of potential energy within a battery is related to the number of volts it has.

What is voltage?

We can think of **voltage** as a measure of the size of 'push' that causes a current to flow around a circuit. Because the current is a flow of charge, something is needed to make the charges move.

If there is no voltage, then there can be no current flowing because there is nothing to cause the charges to move. The larger the voltage, the bigger the 'push' and the more current that can potentially flow.

The symbol for voltage is *V* and the unit is **volts** (V).

The energy source for the voltage is usually a battery or cell, but it can also come from a mains socket. A large energy source, like a big car battery of 12 V, will provide more 'push' or voltage and hence more current than a small cell of 1.5 V. Voltage is measured using a **voltmeter** (Figure 1.2.2a).

If two cells or more are connected together side-by-side, the voltage across them is the sum of the voltage of each cell. This is because both cells are 'pushing' the same way.

> 1. Why does no current flow if there is no voltage?
> 2. Figure 1.2.2b shows two circuits, one with one cell and the other with three cells. If, instead, there were two cells, what reading would the voltmeter give?

Voltage and components

If there is a higher voltage, there will be more current flowing and therefore more energy being transferred to the components. A light bulb will be much brighter if it is connected to a 6 V battery rather than to a 3 V battery in a similar circuit.

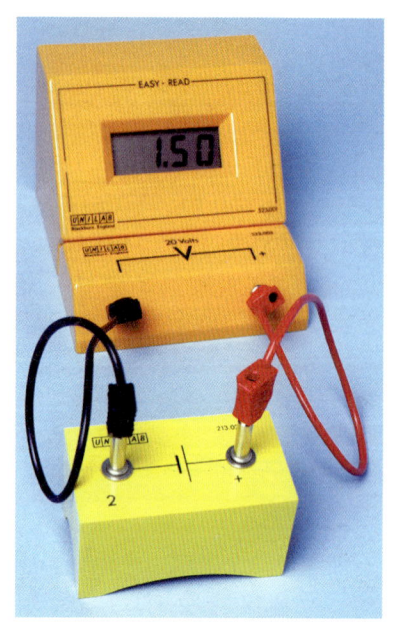

FIGURE 1.2.2a: A voltmeter is a tool to measure voltage, but what do we mean by voltage?

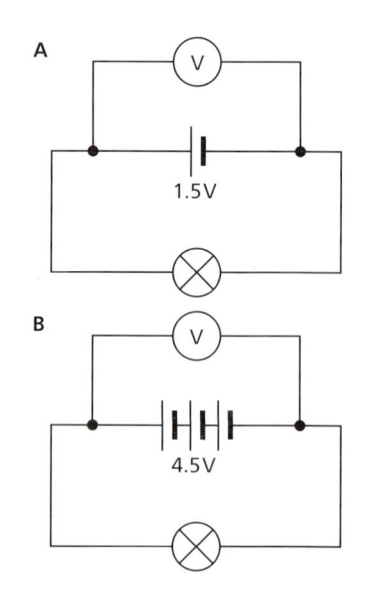

FIGURE 1.2.2b: Measuring the voltage across cells. The circled V represents a voltmeter.

---

34   AQA KS3 Science Student Book Part 1: Electromagnets – Voltage and resistance *and* Current

Figure 1.2.2c shows how the voltmeter must be connected *across* a component (here a bulb) to measure the difference in potential across the component. Voltage is also known as **potential difference**.

> **3.** In which of the circuits in Figure 1.2.2b will the light bulb be the brighter? Explain your answer.
>
> **4.** What might happen to a motor if it were connected to the 230 V mains electric supply rather than to a 12 V battery?

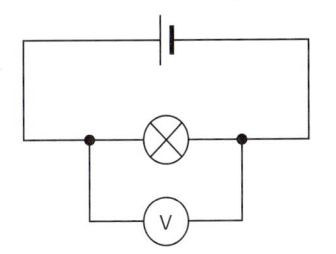

FIGURE 1.2.2c: Measuring the voltage across a bulb.

Using analogies to explain voltage

Imagine blowing gently through a straw. The air flowing through the straw is like a current and the amount of push given to the air is like the voltage. If you blow harder (more voltage) there is more air flow (more current).

A high waterfall is also like a large voltage. It will transfer a lot of energy to the water (charge), making the river flow very fast (a large current). The difference in height makes the river flow. In a circuit, the difference in charge across the battery provides the push for the current.

FIGURE 1.2.2d: The difference in height makes the water move.

Did you know…?

Electric eels can produce electrical discharges of around 500 V in self-defence.

> **5.** Compare a circuit with a 12 V battery and one light bulb with a circuit that has a 1.5 V cell and one light bulb. Use the two analogies above to explain how they will be different.
>
> **6.** Explain one limitation for each of the analogies outlined.

Know this vocabulary

voltage
volt
voltmeter
potential difference

SEARCH: voltage

Electromagnets

Explaining resistance

We are learning how to:
- Explain what resistance is and how it affects the circuit.
- Investigate and identify the relationship between voltage and current.
- Calculate the value of a resistor used in a circuit.

All materials offer some opposition to the flow of current – we call this 'resistance'. The amount of resistance can vary widely, even in different metals. Why are some metals, like gold, better at conducting electricity than other metals, like tin?

What is resistance?

The word 'resistance' means to oppose. In electric circuits, electrical **resistance** opposes the 'push' provided by the voltage. The overall current flowing through the circuit, therefore, depends on both the voltage and the resistance.

If there is a high voltage and a low resistance, then a large current will flow. This is because there is not very much opposition to the 'push' given by the voltage. Imagine a motor in a circuit. The current through it causes it to spin. If the motor is swapped with one of higher resistance, there will be more opposition to the flow of charge and, for the same voltage, the current will be smaller. The motor with a higher resistance will spin more slowly.

All components in a circuit provide some resistance.

> **1.** A buzzer is an electrical device that creates a sound when there is a current through it.
>   **a)** A circuit, A, has a 6V battery and a buzzer. Another circuit, B, has a 6V battery and a buzzer with higher resistance. In which circuit will the buzzer be louder?
>   **b)** Explain your answer to a) using ideas about resistance and current.

Conductors and insulators

Resistance depends on the type of material an object is made from. Materials that are very good conductors of electric current have a very low resistance. Electrical insulators have a very high resistance, and do not allow current to flow easily.

All metals conduct electricity well because they have many **free electrons** that can move when a voltage is applied.

Did you know...?

There are many different models used to help explain electricity. One of them compares a circuit to water being pumped around pipes. If the pipe is narrower the resistance to flow is greater.

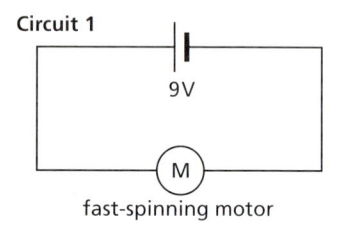

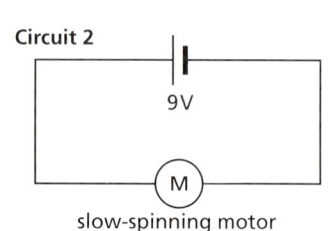

Circuit 1

9V

fast-spinning motor

Circuit 2

9V

slow-spinning motor

FIGURE 1.2.3a: The resistance in circuit 1 is low, so there is a big current; what can you say about circuit 2?

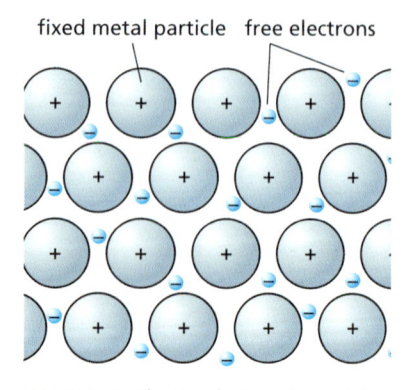

FIGURE 1.2.3b: Conduction in metals depends on free electrons.

2.3

As the electrons move, they will collide with other particles in the metal structure. This is the cause of resistance in most ordinary metals. It is why even the best electrical conductors, like platinum, will have some resistance.

In an insulator, the electrons are more tightly bound to atoms than in a conductor; far fewer electrons are free to move and so there is insignificant current.

2. As an analogy of a circuit with resistance, think of an obstacle race. Which parts of a circuit do the obstacles represent? Which parts of the circuit do the people represent?

3. What would happen to a light bulb if the copper wires in a circuit were replaced with platinum? Explain your answer.

Working out resistance

Resistance is measured with the unit **ohms** (Ω) and is represented by R. All the components in a circuit will have their own resistance. It is possible to investigate the relationship between voltage (V) across and current (I) through a component, as shown in Figure 1.2.3c.

The definition of resistance is:

$$\text{resistance} = \frac{\text{voltage}}{\text{current}} \qquad R = \frac{V}{I}$$

As resistance can be calculated from potential difference and current, we can use this to find out what the resistance of a component is. A team of students is investigating this to see if the resistance of a component stays the same.

Figure 1.2.3d shows the circuit they set up. The rectangle is the resistance – that's what they're trying to find the value of. They altered the settings on the power pack so that there was a range of voltages. Their results are shown in Table 1.2.3.

TABLE 1.2.3: Investigation results.

Potential difference/V	Current/A
0.9	0.03
1.9	0.07
3.1	0.10
3.9	0.12
5.0	0.15
6.1	0.19

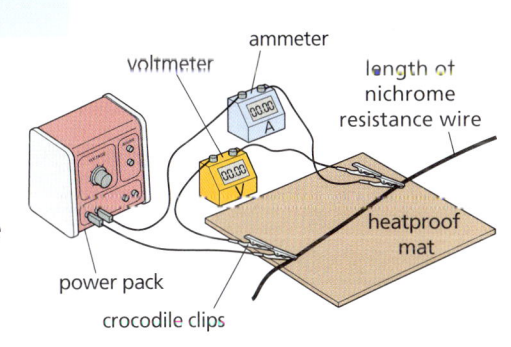

FIGURE 1.2.3c: As the voltage supplied is changed using the power pack, the current is measured using the ammeter. The resistance of the length of nichrome wire between the crocodile clips can then be determined.

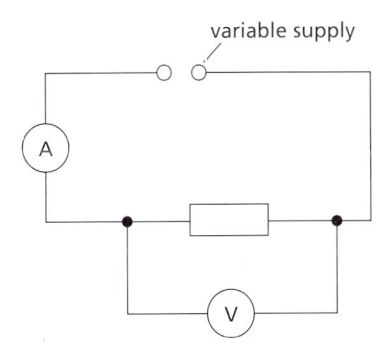

FIGURE 1.2.3d: Circuit to determine resistance.

4. What calculation needs to be done on each pair of readings to find the resistance?

5. a) Calculate the resistance for each pair of readings.

   b) What do you notice about the values?

Know this vocabulary

resistance
free electron
ohm

SEARCH: resistance

Electromagnets

Describing series and parallel circuits

We are learning how to:
- Describe how voltage, current and resistance are related in different circuits.
- Understand the differences between series and parallel circuits.

You have learned about what voltage, current and resistance are. Now you will see how they interact in a circuit.

Relating voltage, current and resistance

The size of the voltage and the size of the resistance both determine how much current flows. Look at the three different circuits in Figure 1.2.4a. In circuit 1, there is a voltage of 3 V and one light bulb of resistance 3 Ω.

In circuit 2, there are two identical light bulbs in series, providing twice as much resistance, but supplied with the same voltage as in circuit 1. The current flowing through the circuit is now less, because there is the same 'push' (voltage) but twice the opposition to the flow of electrons (resistance). The light bulbs are not as bright as in circuit 1.

In circuit 3, there are now two cells and the same two light bulbs, each with a resistance of 3 Ω. The light bulbs will both be just as bright as in circuit 1. This is because the resistance and the voltage are both doubled compared to circuit 1, so the current will be the same.

1. What is the voltage and the resistance of the circuit in Figure 1.2.4b?
2. Explain whether the light bulbs in Figure 1.2.4b are dimmer or brighter than in:
   a) circuit 1; b) circuit 2; c) circuit 3 of Figure 1.2.4a.

Series and parallel circuits

In a **series circuit**:
- All the components are connected, one after the other, in a complete loop of conducting wire.
- There is only one path that the current can take.
- The voltage is shared between the components.

Figure 1.2.4c shows a series circuit with two light bulbs.

In a **parallel circuit**:
- Each component is connected separately in its own loop between the two terminals of a cell or battery.

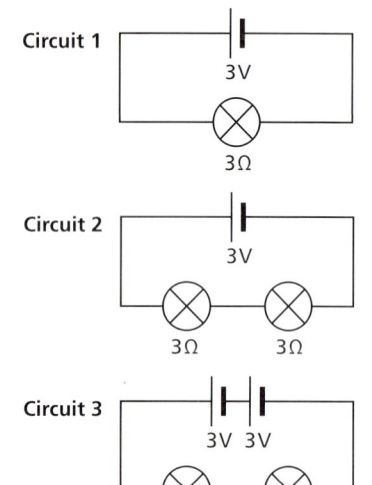

FIGURE 1.2.4a: How does voltage and resistance change in these circuits?

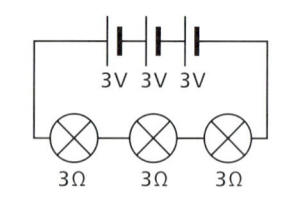

FIGURE 1.2.4b

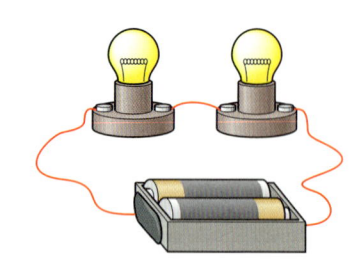

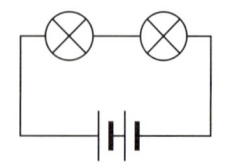

FIGURE 1.2.4c: How can you tell that the components in this circuit are connected in series?

- The full voltage is supplied to each loop.
- The current from the battery is divided between the loops.

Figure 1.2.4d shows a parallel circuit with two light bulbs.

A parallel circuit is rather like separate series circuits connected to the same energy source. The different components are connected by different wires. Therefore, if a bulb blows or is disconnected from one parallel wire, the components in the other loops keep working because they are still connected to the battery in a complete circuit.

If more bulbs are added in parallel, all the bulbs light up with the same brightness as before, because the potential difference across each is the same (equal to the battery voltage).

3. What would happen to the components in a series circuit if one of the bulbs stopped working?

4. Draw two circuits – one with just one bulb, and the other with three identical bulbs in series. Both circuits should have just one cell of the same voltage. Compare:
   a) the voltage in each circuit;
   b) the current in each circuit;
   c) the brightness of the bulbs in each circuit.

5. a) Draw a parallel circuit with four bulbs.
   b) Explain how this is different from a series circuit with four bulbs.

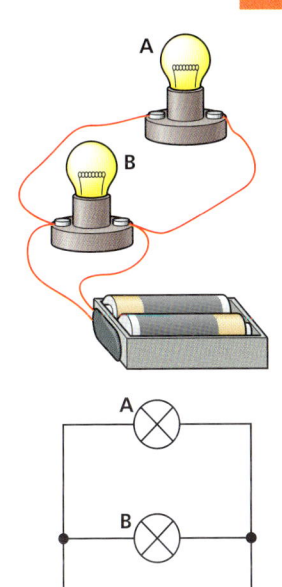

FIGURE 1.2.4d: What happens to bulb A in this parallel circuit if bulb B 'blows'?

Explaining series and parallel circuits

When two light bulbs are connected in series, the resistance in the circuit is increased compared to that with one light bulb. The increased resistance opposes the flow of current, so fewer electrons pass per second, transferring less energy. The light bulbs are therefore not as bright as in a circuit with the same voltage but only one bulb.

However, when two light bulbs are connected in parallel, each loop behaves like a separate circuit. The resistance in each branch is the same as if there were just one light bulb in the whole circuit. There is the same current in each branch of the circuit, so the bulbs light up with the same brightness as in the single-bulb circuit. The energy stored in the battery will decrease twice as quickly and the battery will run out faster than in a series circuit.

6. Explain the advantages and disadvantages of arranging components in series or in parallel.

Did you know…?

Most circuits used are combinations of series and parallel parts.

FIGURE 1.2.4e

Know this vocabulary

series circuit
parallel circuit

SEARCH: series and parallel circuits 39

Electromagnets

Comparing series and parallel circuits

We are learning how to:
- Investigate and explain current and voltage in series and parallel circuits.
- Explain the circuits in our homes.

The arrangement of components in either series or parallel affects the amount of voltage they receive and the amount of current flowing through them. Why does the arrangement make this difference?

Current and voltage in series and parallel circuits

Figures 1.2.5a and 1.2.5b show a series circuit and a parallel circuit with light bulbs of the same resistance.

Series circuit

The ammeter shows the same readings in different parts of the circuit.

However, the voltage is divided between the components. See how the voltage across each of the components adds up to the total provided. We can write this as:

$$V_{total} = V_1 + V_2 + V_3$$

If the components have the same resistance, the voltage is divided equally.

Parallel circuit

The voltage in all parts of the circuit is the same regardless of how many loops there are.

However, the current splits up between each loop. Adding up the current in each part gives the total current flowing from the battery. We can write this as:

$$I_{total} = I_1 + I_2 + I_3$$

If the resistance in each part is the same, the same current will flow through each.

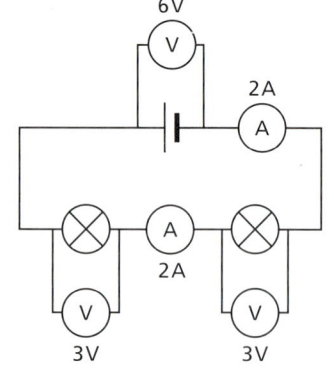

FIGURE 1.2.5a: A series circuit.

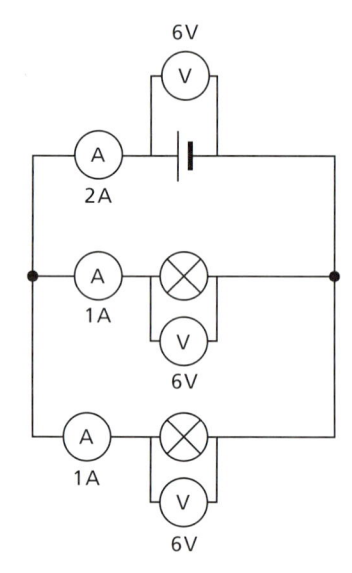

FIGURE 1.2.5b: A parallel circuit.

1. If another light bulb is added to the series circuit in Figure 1.2.5a, what will happen to the voltage across the other light bulbs? Explain your answer.

2. A 12 V battery is connected in a circuit with ten identical light bulbs in parallel. Compare this with the circuit in Figure 1.2.5b. What will the current be in each individual loop?

Did you know...?

During World War 2 there was a shortage of copper. In 1942, the ring main helped to reduce the amount of household wiring needed. This required longer wires, but they could be thinner.

2.5 Selecting series or parallel circuits, according to application

Each type of circuit has advantages and disadvantages. These are summarised in Table 1.2.5.

	Advantages	Disadvantages
Series circuits	Simple to set up. Shares the voltage between the components, which is useful if the components need a lower voltage than the supply voltage.	If one component fails, the whole circuit stops working. If more components are added, each gets less voltage and so might not work as well.
Parallel circuits	If a component in one of the loops fails, the other loops keep functioning. Each loop gets the same voltage, so adding more loops doesn't mean other loops suffer a voltage drop.	More complicated to set up. Adding more loops doesn't reduce the voltage, so if components need a lower voltage they won't work.

TABLE 1.2.5

We can therefore decide which type of circuit to use in each situation.

3. Which kind of circuit would you use to supply 20 bulbs, each rated at 12 V, with a power of 240 V?
4. Which kind of circuit would you use for emergency lighting in a restaurant, running from a 12 V supply and powering 12 V bulbs?

Did you know...?

A wiring system that consists of wires arranged in complete loops around a building is called a ring main.

Mains supply is a supply of electricity to a building at the standard voltage for that area (230 V in the UK).

Household circuits

Figure 1.2.5c shows how the household electricity supply is connected in the UK. It is an arrangement known as the domestic **ring main**.

All the plug sockets in the ring main are connected in parallel, for the following reasons:

- If one of the electrical appliances should stop working, other appliances are not affected.
- The **mains supply** of 230 V is applied across all the sockets.
- Switches can be used to turn the current on and off within each branch.

FIGURE 1.2.5c Arrangement of sockets in a domestic ring main.

FIGURE 1.2.5d Each socket has 230 V applied to it.

5. Suggest disadvantages with this arrangement.

Know this vocabulary

ring main
mains supply

SEARCH: current and voltage in series and parallel circuits   41

Electromagnets

Investigating static charge

We are learning how to:
- Recognise the effects of static charge.
- Explain how static charge can be generated.
- Use evidence to develop ideas about static charge.

Static electricity is a common and sometimes spectacular phenomenon. You may have noticed that after walking across a carpet, you sometimes get a small electric shock when you touch a door handle. This happens when your body has become electrically charged. Lightning is a demonstration of static electricity at work on a grand scale.

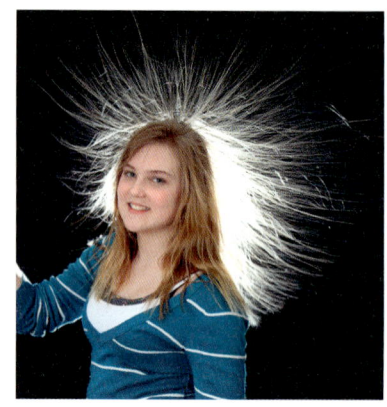

FIGURE 1.2.6a: When a person's hair becomes charged the individual strands repel each other.

Static charge

Electric **charge** can either flow or be gathered in one place. Charge that is flowing is called a current and when it is not flowing it is called **static electricity**.

Electricity flows through conductors, such as a copper wire. However, when a charged material is not connected to a conductor, the electricity cannot flow away and so the charge stays in place.

When a charged object comes close to a conductor the electricity jumps across as a spark. If your body has become charged by walking on a carpet, you feel the charge flowing away through your fingers when you reach for the door handle.

1. Name some materials that are good conductors of electricity.
2. What does the word 'static' mean?
3. How could a material that conducts electricity become charged?

Attraction and repulsion

When an object becomes charged with static electricity, a **field** of electrostatic force exists around the object. This is a non-contact force. This force can **attract** other materials and may be strong enough to lift them. A charged balloon brought close to someone's head can attract strands of hair and lift them up without the balloon coming into contact with the hair. Scraps of paper can be made to jump off a table and stick to a charged plastic comb held a few centimetres above it.

FIGURE 1.2.6b: Static electricity can cause attraction.

42 AQA KS3 Science Student Book Part 1: Electromagnets – Voltage and resistance *and* Current

When two objects of the same material become charged they **repel** each other.

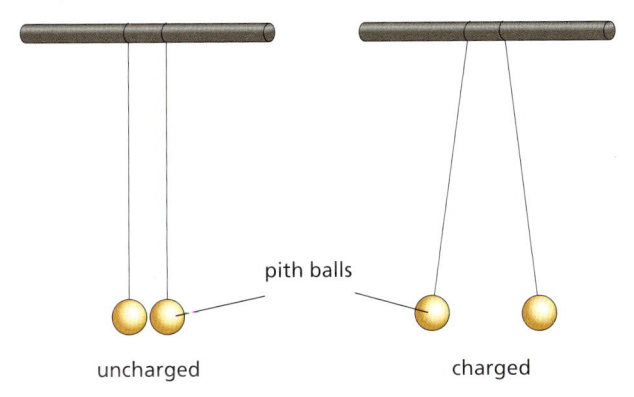

pith balls

uncharged   charged

FIGURE 1.2.6c: Repulsion between two identical charged objects.

4. What could small pieces of dust and paper experience when a charged object is brought close?
5. How could you find out if two charged combs repel each other?

Factors affecting field strength

If you push a supermarket trolley you are applying a **contact force** because you are in contact with the object. Electrostatic force, however, like gravity, is a **non-contact force**. If we have an object with a static charge it produces an electric field. Fields are an important idea in science.

A charged balloon will produce an electric field – an area around the balloon in which anything affected by the charge will be subject to a force. If we were to use the charged balloon to attract bits of tissue paper, it would work only if the bits were within the field.

The strength of the field will depend on two things:
- the strength of the charge on the balloon;
- the distance between the balloon and the bits of paper.

6. What evidence supports the idea that static electricity exerts a non-contact force?
7. How could you show that the field around the balloon got weaker as the distance from the balloon increased?
8. There's another non-contact force we've met already, as well as electrostatic. What is it?
9. Some people don't like getting shocks from static charge. Suggest how they could reduce the likelihood of getting them.

Did you know...?

Some items of clothing become charged so easily that when you take them off, the cloth crackles and sparks as the charge escapes. This occurs in dry weather, and a dark room is needed to see the effect.

Know this vocabulary

charge
static electricity
field
attract
repel
contact force
non-contact force

SEARCH: static electricity

Electromagnets

Explaining static charge

We are learning how to:
- Explain static charge in terms of electron transfer.
- Apply this explanation to various examples.

In ancient Greece, people started to put forward ideas about atoms. They thought that atoms were the most basic particles and that they could not be split further. It was not until the 1800s that ideas really developed beyond this. Scientists have developed a much better understanding of what atoms are like inside. These more modern ideas form the basis of our understanding in many areas of chemistry and physics, including static electricity.

Atoms and electrons

The simplest modern model of an atom is a nucleus being orbited by **electrons**. The nucleus has a positive electric charge because it contains positively charged **protons** – along with neutrons, which have no charge. Electrons have a negative electric charge. Overall, an atom is electrically neutral because the positively charged protons are balanced by an equal number of negatively charged electrons.

If some electrons get transferred from one object to another the charges no longer balance. This is what happens when an object becomes **charged up**.

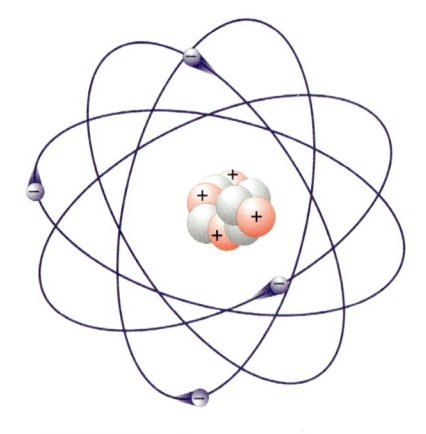

FIGURE 1.2.7a: Atoms contain a balance of positively charged protons and negatively charged electrons.

1. What are atoms made up of?
2. Why do atoms have no charge overall?
3. How can an object become negatively charged?
4. How can an object become positively charged?

Positive and negative charge

When a nylon rod is rubbed with a cloth, electrons are transferred from the rod to the cloth. Because electrons have negative charge this makes the cloth **negatively charged**. The rod has lost electrons so the positive charge of the protons is no longer balanced – the rod is left **positively charged**.

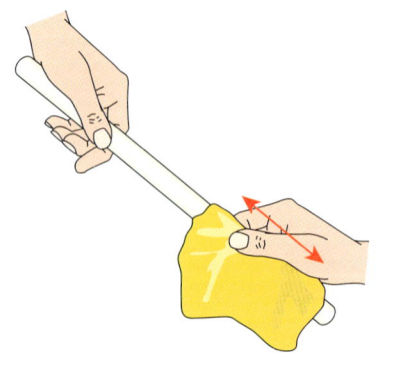

FIGURE 1.2.7b: Rubbing transfers electrons, either from the rod to the cloth or from the cloth to the rod.

44   AQA KS3 Science Student Book Part 1: Electromagnets – Voltage and resistance *and* Current

Other materials behave differently. A polythene rod, for example, gains electrons when rubbed with a cloth. It becomes negatively charged and the cloth, which has lost electrons, becomes positively charged.

> 5. Describe what happens to a cloth when it is rubbed on a nylon rod.
> 6. Explain how different materials behave differently when rubbed with a cloth.

Charging by electron transfer

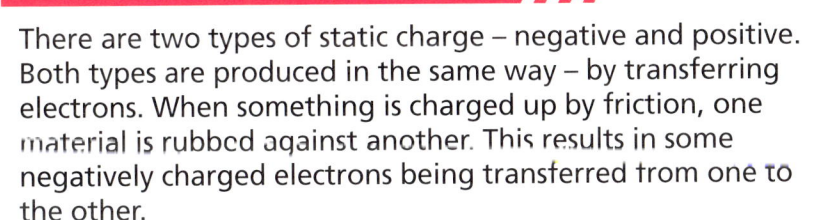

There are two types of static charge – negative and positive. Both types are produced in the same way – by transferring electrons. When something is charged up by friction, one material is rubbed against another. This results in some negatively charged electrons being transferred from one to the other.

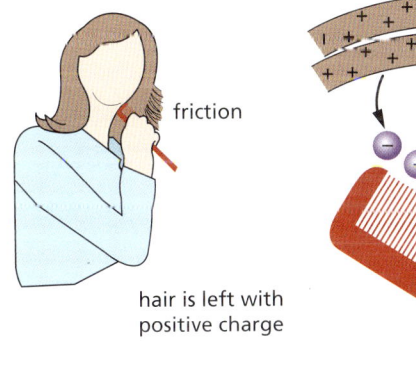

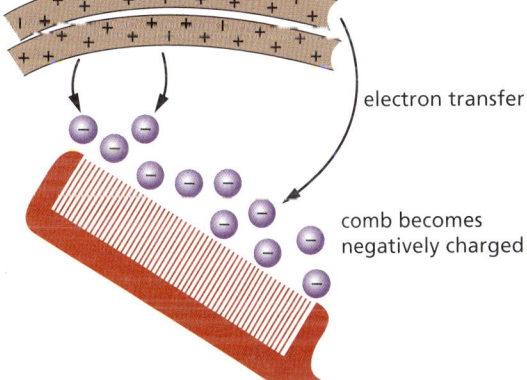

This means that one material has lost electrons and doesn't have enough to balance out its positively charged protons. This material now has a positive charge. The other material has gained excess electrons so it has a negative charge.

For example, if you brush your hair vigorously with some types of plastic comb or brush, your hair becomes charged and so does the comb. In this case, electrons are being transferred from your hair to the comb. This means your hair is lacking electrons and has a positive charge. The comb has gained electrons and has a negative charge.

> 7. How can you tell that your hair has become charged?
> 8. How could you show that the comb has also become charged?
> 9. Neither your hair nor the comb will stay charged permanently. Using the idea of electron transfer, suggest what you think happens.

2.7

Did you know…?

A Van de Graaff generator produces electricity by friction. The ones used in schools can produce 100 000 volts. Bigger Van de Graaff generators can exceed two million volts.

FIGURE 1.2.7c: Electrons are transferred when you comb your hair with a plastic comb.

Know this vocabulary

electron
proton
charged up
negatively charged
positively charged

SEARCH: electron transfer in static charge   45

Electromagnets

Understanding electric fields

We are learning how to:
- Explain static electricity in terms of fields.
- Explain how charged objects affect each other.

Charged objects can affect their surroundings even when they are not in contact. Sometimes people believe that they can 'feel' electricity in the air. The idea of an electric field helps to explain this.

Rules of attraction and repulsion

An **electric field** exists around a charged object, which can exert a non-contact force.

Two similarly charged objects – both negative or both positive – **repel** each other. This is called repulsion.

Two oppositely charged objects – negative and positive – **attract** each other.

A charged object can also attract an uncharged object. For example, water has no overall electrical charge – it has a balance of negatively and positively charged particles. Despite this, water is affected by an electric field (Figure 1.2.8a).

1. What is the area around a charged object called?
2. Looking at Figure 1.2.8a, what evidence suggests that a non-contact force is working?
3. All substances contain charged particles. But most objects have no charge – explain why.

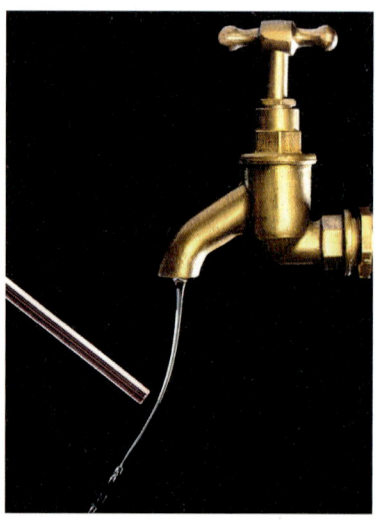

FIGURE 1.2.8a: The water is attracted towards a charged rod.

Charged particles moving

Within many substances, charged particles are free to move. When there is no electric field present, the charged particles are spread evenly.

In Figure 1.2.8b the negatively charged balloon has electrons spread over its surface. When it is brought towards the wall, the negatively charged particles in the wall atoms are repelled. This leaves the surface of the wall with a positive charge. The opposite charges of the balloon and the wall's surface attract one another.

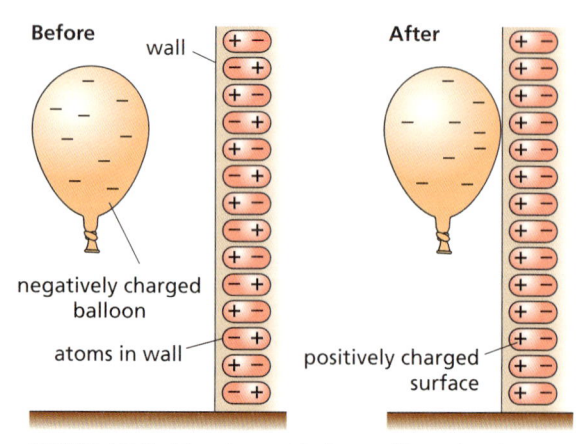

FIGURE 1.2.8b: The charged balloon affects the charges in the atoms of the wall.

46 AQA KS3 Science Student Book Part 1: Electromagnets – Voltage and resistance *and* Current

2.8

4. Describe how charged particles move when an object is put in an electric field.
5. Draw a labelled diagram, similar to Figure 1.2.8b, to show how a positively charged rod can attract a trickle of water.
6. Suggest why a metal rod is unlikely to be able to attract a trickle of water.

Loss of charge

Static charge depends on electrons being unable to flow into or out of an object. If a charged polythene rod is connected to a conductor, such as a wire, electrons will flow away from the rod. The rod loses its charge and becomes neutral.

Air is not a good conductor, but it can transfer some electrons, so charged objects gradually lose their charge. In wet weather, the water vapour in the air can transfer more electrons so charge is lost more quickly.

When a Van de Graaff generator is turned on, the globe becomes positively charged. If the charge builds up enough, the air can start to conduct. Sparks will jump across the gap to anything in good contact with the ground.

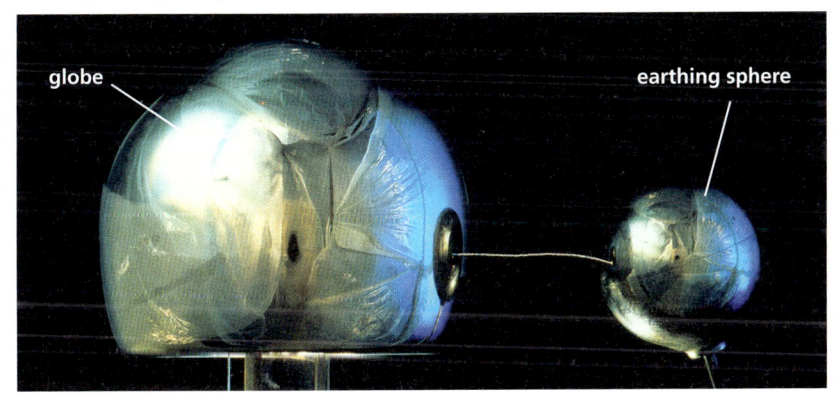

FIGURE 1.2.8c: A Van de Graaff generator and earthing sphere.

Did you know...?

An electrostatic field exists around a charged object in three dimensions – above and below it as well as on all sides.

7. Explain why experiments with static electricity give better effects in dry weather.
8. a) Explain why, if someone is charged up by a Van de Graaff generator, their hairs rise up and spread out.
   b) Explain the process of discharging the globe of a Van de Graaff generator.
9. Suggest ways of avoiding getting electrostatic shocks in everyday life.

Know this vocabulary

electric field
repel
attract

SEARCH: Van de Graaff generator   47

Electromagnets

Checking your progress

To make good progress in understanding science you need to focus on these ideas and skills.

☐ Recognise arrangements of electric circuit components in series and in parallel.	☐ Use circuit diagrams to construct real series and parallel circuits and vice versa.	☐ Suggest the advantages of series and parallel circuits for particular applications.
☐ Describe what is meant by current, voltage and resistance.	☐ Apply a range of models and analogies to describe current, voltage and resistance.	☐ Evaluate different models and analogies for explaining current, voltage and resistance.
☐ Know that a complete circuit is needed for current to flow.	☐ Know that current is a movement of electrons and is therefore a flow of charge.	☐ Know that current is divided between the loops in a parallel circuit.
☐ Know that resistance reduces the current flowing.	☐ Explain the idea of resistance, using models such as water flow in pipes.	☐ Understand that resistance is the ratio of voltage to current.
☐ Understand that voltage is also called potential difference and this makes current flow around a circuit.	☐ Understand that in a series circuit the potential difference is shared by the components.	☐ Understand that potential difference is the amount of energy transferred from the battery to the charge or from the charge to the components.

2.9

- ☐ Describe the relationship between current, voltage and resistance in a qualitative way.
- ☐ Use data to identify a pattern between current, voltage and resistance.
- ☐ Use data and the mathematical relationship between current, voltage and resistance to carry out calculations.

- ☐ Describe the effect that a charged object has on other charged objects.
- ☐ Explain what is meant by an electrostatic field.
- ☐ Suggest how objects may become electrostatically charged.

- ☐ Know the two types of static charge.
- ☐ Explain how electron transfer can result in either type of charge.
- ☐ Explain the operation of a circuit using the idea of electrons moving from the negative to the positive terminals of a power supply.

- ☐ Describe how friction between objects may cause electrostatic charge through the transfer of electrons.
- ☐ Explain various examples of electrostatic charge; use ideas of electron transfer to explain different effects.
- ☐ Explain why some electrostatic charge mechanisms are more effective than others.

Electromagnets

Questions

KNOW. Questions 1–3

See how well you have understood the ideas in this chapter.

1. What is the unit of current? [1]
   a) volt   b) ohm   c) amp   d) joule

2. Explain how a series circuit is different from a parallel circuit. [2]

3. Thinking about electrostatic charge, which of these statements is true? [1]
   a) positive (+) charge repels negative (−) charge
   b) positive (+) charge attracts positive (+) charge
   c) negative (−) charge attracts positive (+) charge
   d) negative (−) charge attracts negative (−) charge.

APPLY. Questions 4–6

See how well you can apply the ideas in this chapter to new situations.

4. Figure 1.2.10a shows four circuits A–D. Which of the following shows the correct order, from the circuit that has the brightest bulbs to the one that has the dimmest? [1]
   a) A, B, C, D   b) D, C, B, A   c) C, D, A, B   d) C, B, D, A

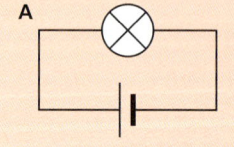

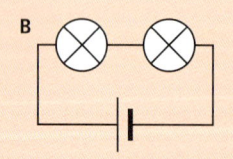

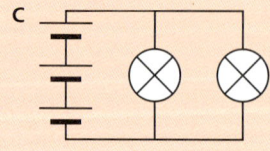

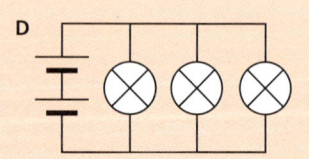

   FIGURE 1.2.10a: All cells and all lamps are identical.

5. Figure 1.2.10b shows a model of a circuit. How would you change this model to show an increased voltage and increased resistance? [2]

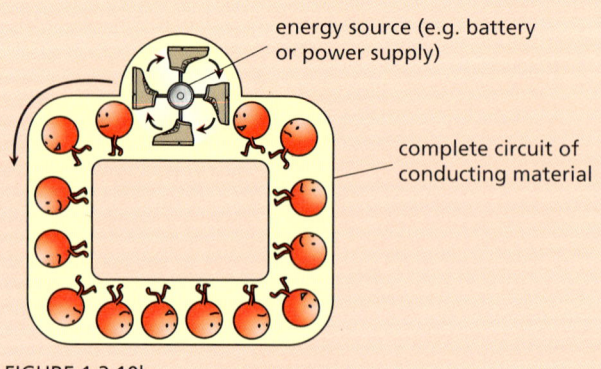

   FIGURE 1.2.10b

6. Suggest how you could find out if one electrostatically charged rod has more charge than another of the same material. [2]

50 AQA KS3 Science Student Book Part 1: Electromagnets – Voltage and resistance *and* Current

2.10

EXTEND. Questions 7–8

See how well you can understand and explain new ideas and evidence.

7. Table 1.2.10 gives some data from an investigation comparing the different lengths of the same wire. The values of resistance have been calculated using $V/I = R$.

   Plot a graph of the resistance against the length of the wire.

   TABLE 1.2.10

Length of wire (cm)	Average voltage (V)	Average current (A)	Average resistance (Ω)
10	0.47	0.23	2.0
20	0.59	0.17	3.5
30	0.64	0.13	4.9
40	0.69	0.11	6.3
50	0.72	0.09	8.0
60	0.76	0.07	10.9
70	0.82	0.06	13.7

8. Jo is asked to construct a circuit with a battery and two parallel loops, each containing two bulbs in series. [4]

   a) Draw the circuit diagram.
   b) If the total resistance in the circuit is 20 ohms and the voltage supplied by the battery is 5 V, how much current will flow out of the battery?
   c) Show on your diagram where an ammeter could be put in the circuit to check this.
   d) Explain what will happen to the other bulbs if one of the bulbs should blow.

Energy
Energy costs *and* Energy transfer

Ideas you have met before

Although you'll have heard and used the term 'energy' before, you may well not have explored its use in science. However, there are many things you will have investigated which use energy.

Materials may change
Changes such as burning result in the formation of new materials.

Some materials change state when they are heated or cooled.

Living things need nutrition
Plants require light and water for life and growth.

Animals need nutrition, and they cannot make their own food; they get nutrition from what they eat.

Food chains identify producers, predators and prey.

Nutritional Information per 100 g	
Protein	18.1 g
Fat	16.2 g
Of Which Saturates	(5.2 g)
Carbohydrates	26 g

Objects can move in various ways
Unsupported objects fall towards Earth because of gravity.

Air resistance, water resistance and friction act between moving surfaces.

Light and sound travel as waves
We see things because light travels from light sources to our eyes, or from light sources to objects and then to our eyes.

Sounds are made because of something vibrating; these vibrations travel through the air to our eyes.

Electricity can do useful work
Many common appliances run on electricity.

Components in a circuit can be made to function in different ways, for example lamps can be made brighter and buzzers can be made louder.

In this chapter you will find out

3.0

Energy stores and transfers
- Energy is transferred when changes happen, and this transfer can happen in many different ways.
- An object stores energy if it has been raised up. This is because it is affected by the Earth's gravitational force.
- When elastic materials are stretched or squashed they have more energy stored in them.

Fuels are energy stores
- Fuels are energy stored chemically. They include wood, fossil fuels and hydrogen.
- Fuels only burn if oxygen is present. The products of burning also store energy, but less than that in the fuel and oxygen.
- When a fuel is burned in oxygen, energy is transferred to the surroundings.

Energy in the home
- The quantity of energy transferred in a change can be measured.
- How quickly energy is transferred is the power and this can also be measured.
- Electricity is generated by using different energy resources, which each have advantages and disadvantages.
- We pay for our domestic electricity based on the amount of energy transferred.
- We can calculate the cost of home energy usage using the formula: cost = power (kW) × time (hours) × price (per kWh).

Accounting for energy
- We can describe how jobs get done using an energy model where energy is transferred from one store at the start to another store at the end.
- When energy is transferred, the energy total is conserved, but some energy is dissipated, reducing the useful energy.

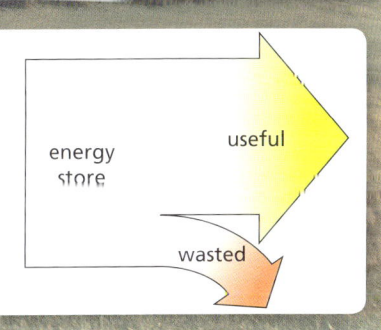

energy store → useful / wasted

53

Energy

Understanding energy transfer by fuels and food

We are learning how to:
- Describe the use of fuels in the home.
- Explain that foods are energy stores and that the amount stored can be measured.
- Explain that energy is not a material and can be neither created nor destroyed.

Natural gas and electricity are used in homes to supply energy. Our bodies, too, need supplies of energy. But is energy for the body the same as energy for the home?

Fuels and energy in the home

One of the places we use energy is in our homes. We need light and heat and we need to power appliances such as TVs and washing machines.

Fuels are a way of providing us with the energy we need.

FIGURE 1.3.1a: Gas pipelines and electricity cables supply us with energy.

1. Name four fuels that might be used in homes.
2. Describe how energy use in a home is measured.
3. Suggest an advantage and a disadvantage for these fuels:
   a) electricity;
   b) gas.

Food and energy

Energy is measured in **joules** (abbreviated to J). A joule is a very small amount of energy so we sometimes use **kilojoules (kJ)**: 1000 J = 1 kJ. You may hear people talk about calories – this is an older unit of energy that scientists no longer use.

Food is fuel for our bodies. Energy stored in food is often called its 'energy content', measured in kilojoules (kJ); 1 kJ = 1000 J and this is equivalent to 240 'calories'. During digestion food is changed into chemicals that store energy in the body's cells (an **energy resource**).

Did you know...?

The unit of energy, the joule, is named after James Joule, a Salford brewer who discovered the link between heat and work. He spent part of his honeymoon in Switzerland making measurements to show that the water at the bottom of a waterfall was slightly warmer than the water at the top.

3.1

Chemical reactions that happen in the body enable growth and reproduction, responses to the environment and keeping healthy. All rely on energy being transferred from chemical stores in the body's cells which, in turn, depends on the food eaten.

> 4. Explain how the body builds up stores of energy.
> 5. Explain why information about energy stored in food is useful.
> 6. Calculate the energy content of 100 g of the food product shown in Figure 1.3.1b.

Transfers and stores

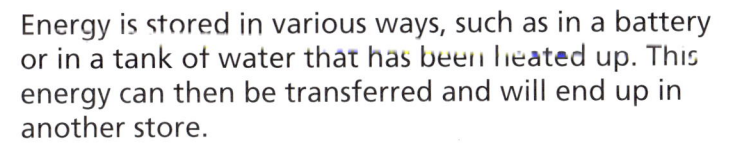

Energy is stored in various ways, such as in a battery or in a tank of water that has been heated up. This energy can then be transferred and will end up in another store.

For example, when a fuel burns in air, energy stored in the fuel and in oxygen is transferred to the surroundings, which warm up. The energy stored in the products of combustion and the warmer surroundings equals the energy stored in the fuel and oxygen.

Similarly, warming a room with an electric heater causes a change that results in energy being transferred to the surroundings (air, walls, ceiling, furniture and so on). The total amount of energy remains the same, even though it is more spread out.

When we eat food, energy stored in food is transferred to energy stored in our bodies. This stored energy is transferred further during body processes.

When these changes happen, energy is not used up but is transferred to different places.

> 7. Give two other examples of changes taking place that involve energy transfer, and explain where you think the energy has been transferred to.
> 8. When a candle is burning:
>   a) How is energy being transferred?
>   b) Where is it being transferred from and to?

Nutritional Information per 100 g	
Protein	18.1 g
Fat	16.2 g
Of Which Saturates	(5.2 g)
Carbohydrates	26 g
Of Which Sugars	(7.2 g)
Sodium	0.468 g
Potassium	906 mg
Salt	1.2 g
Fibre	4.7 g
kCalories	322 kCal
kJoules	1347 kJ

FIGURE 1.3.1b: Energy content is given as part of the nutritional data on a food label. This shows what is contained in a 100 g serving of food.

Know this vocabulary

fuel
joule
kilojoule (kJ)
energy resource

SEARCH: food, fuel, energy resource

Energy

Comparing rates of energy transfers

We are learning how to:
- Describe what is meant by 'rate of energy transfer'.
- Recall the correct units for rate of energy transfer.
- Calculate quantities of energy transferred.

Chinese food is often cooked in a wok. Using a frying pan instead never seems to produce the same flavours. The reason is speed of cooking – woks are thinner than frying pans (about one-third the thickness). Energy is transferred much more quickly through them and the food is cooked more quickly – essential to create that authentic Chinese taste. How do we measure how quickly energy is transferred?

FIGURE 1.3.2a: The design of a wok speeds up the energy transfer to the food.

Energy and power

Sometimes energy is transferred very quickly and other times it takes much longer. When we talk about how quickly energy is transferred we use the word '**power**'.

Power is measured in **watts** (W). Transferring one joule every second would mean 1 W of power. 1 W is a small amount of power so we often use **kilowatts** (kW): 1 kW = 1000 W.

If the change can be controlled, so can the rate at which energy is transferred. For example, to make a lamp transfer energy more quickly it would need a more powerful light bulb.

FIGURE 1.3.2b: Warm clothes slow the rate of energy transfer from the children's bodies to the cold air, helping to keep them warm.

1. How many watts are there in 2 kW?
2. Of the appliances shown in Figure 1.3.2c:
   a) Which has the highest power rating?
   b) Which has the lowest power rating?
3. How many joules of energy does a 100 W bulb transfer every second?

electric kettle, 2000 W (2 kW)

electric oven, 2150 W (2.15 kW)

laptop computer, 20 W

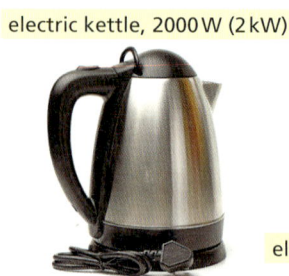

microwave oven, 1000 W (1 kW)

toaster, 1200 W (1.2 kW)

FIGURE 1.3.2c: Some typical power ratings.

Calculating power   3.2

We can calculate power by using the formula:

$$\text{power} = \frac{\text{energy transferred}}{\text{time taken for transfer}}$$

For example, if a light bulb transfers 1000 J of energy in 10 seconds, the power is:

$$\frac{1000}{10} = 100\,\text{W}$$

When we talk about how quickly something happens, we sometimes refer to the rate. This is the amount per second.

4. Why does a toaster have a much higher power output than a laptop computer?
5. What is the power of a bulb that transfers 600 J per minute?
6. What is the rate of energy transfer in joules per second for a 20 W laptop?

FIGURE 1.3.2d: The power rating of the electric fan heater is shown on the label. It is 2000 W.

Quantities of energy transferred

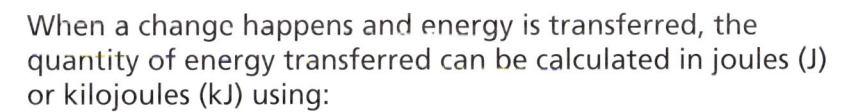

When a change happens and energy is transferred, the quantity of energy transferred can be calculated in joules (J) or kilojoules (kJ) using:

energy transferred (J or kJ) = power (W or kW) × time (s)

- A 20 W laptop computer transfers 20 J/s.
  So if it is used for one hour (1 × 60 × 60 = 3600 s), it transfers 20 × 3600 = 72 000 J = 72 kJ.
- A 2.15 kW electric oven transfers 2.15 kJ of energy per second.
  So if it is used for one hour (1 × 60 × 60 = 3600 s), it transfers 2.15 × 1000 × 3600 = 7 740 000 J = 7.74 MJ.

7. How much energy is transferred when a 1.2 kW toaster runs for three minutes?
8. Calculate the energy transferred when one store transfers energy to another store by heating it for five minutes at a rate of 15 J/s.
9. a) Calculate the energy transferred when a 10 W bulb is left on for three days.
   b) Calculate the energy saved if the same light bulb is turned off every day for eight hours.

Did you know...?

Some of the ways energy is transferred by a device are more useful than others. Getting light from a bulb is great but heat less so. The more of the energy output that is useful, the more efficient we say the device is.

Know this vocabulary

power
watt
kilowatt

SEARCH: energy transfer and power

Energy

Looking at the cost of energy use in the home

We are learning how to:
- Describe the information a typical fuel bill provides.
- Explain and use the units used on a fuel bill.
- Explain how the costs of energy used can be calculated.

When you look at your gas or electricity bill there are two charges. One is for the amount used, and the other is a fixed charge. Why do energy providers make a fixed charge on top of the cost of the electricity or gas used?

Fuel bills

Electricity and gas for the home are bought from energy suppliers. Users receive energy bills that show:

- the 'standing charge' – a fixed amount regardless of how much energy is used
- the price of each unit of energy and the number of units used.

The amount of electricity used in a home is measured in a unit called **kilowatt-hours (kWh)** by an electricity meter. The quantity of gas is shown on a gas meter in cubic metres (m^3). This will probably be converted into kilowatt-hours (kWh) on an energy bill.

$1 \text{ kWh} = 3\,600\,000 \text{ J}$ or 3600 kJ

A typical home might use several hundred kilowatt-hours (kWh) of energy every month (see Figure 1.3.3a) – in other words, hundreds of millions of joules.

Meter readings (E = estimate, C = customer, A = actual)

Electricity readings

Period	Meter no.	Previous	Present	Rate	kilowatt-hours
4 Sept to 12 Nov	S08B 06654	12549 E	12757 C	Normal	208

Gas readings

Period	Meter no.	Previous	Present	Units	kilowatt-hours
30 Aug to 12 Nov	674215	02938 A	02954 C	16 m^3	converts to 178

Charges

Electricity charges

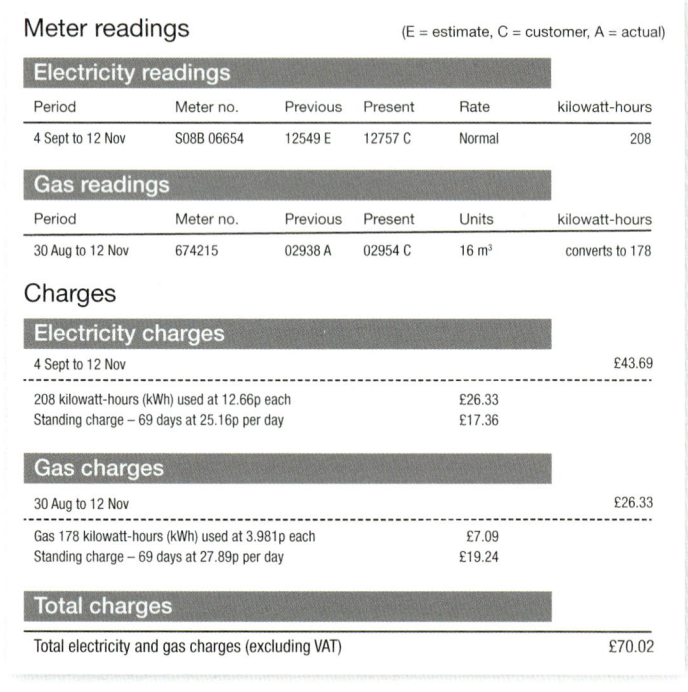

4 Sept to 12 Nov		£43.69
208 kilowatt-hours (kWh) used at 12.66p each	£26.33	
Standing charge – 69 days at 25.16p per day	£17.36	

Gas charges

30 Aug to 12 Nov		£26.33
Gas 178 kilowatt-hours (kWh) used at 3.981p each	£7.09	
Standing charge – 69 days at 27.89p per day	£19.24	

Total charges

Total electricity and gas charges (excluding VAT)	£70.02

FIGURE 1.3.3a: This energy bill shows the quantities and charges for electricity and gas used in a home.

1. Suggest why electricity bills do not show energy usage in joules.

2. What are the standing charges shown in the energy bill in Figure 1.3.3a for the following?
   a) electricity   b) gas

3. Explain the difference between a standing charge and the cost of energy used. (Hint: As well as the fuel used, what else needs to be paid for?)

Did you know...?

About half of the household cost for gas and electricity is for heating the home. Double glazing can reduce heating bills, but it is expensive to install. It has been estimated that it can take 80 years before the savings balance out the cost of the double glazing.

3.3 Calculating the energy used by domestic appliances

Remember that the rate at which energy is transferred is called power, measured in watts (W) – 1 watt = 1 joule per second (1 W = 1 J/s). The amount of energy used by an appliance is calculated by multiplying its power by the time for which it was used. Electricity supply companies use the energy unit kilowatt-hour (kWh), so we need to use hours, not seconds, in the calculation.

> energy used (kWh) = power (kW) × time (h)

An appliance with a power rating of 500 W running for five hours transfers 0.5 × 5 = 2.5 kWh. Choosing an electrical appliance with the optimum power rating for the intended purpose is important. A more powerful electric kettle will use more energy per second but it will take less time to do the job.

4. How much energy would be used by:
   a) A 2 kW oven in 30 minutes?
   b) A 1 kW microwave in 6 minutes?
   c) A 30 W bulb in an hour?
5. Explain the difference between 'energy' and 'power'.

Calculating the cost of energy used

The cost of home energy usage is calculated using the formula:

> cost = energy used (kWh) × price of energy per kWh

The typical price of 1 kWh of electricity is about 13p, and the typical price of 1 kWh of gas is about 4p. If you used 900 kWh of electricity and 700 kWh of gas, the cost would be:

electricity: 900 × 13 = 11 700 p = £117.00

gas: 700 × 4 = 2800 p = £28.00

So the total energy cost = £117.00 + £28.00 = £145.00

6. Alex is working out how much it will cost to cook a frozen curry using a microwave oven compared with a gas cooker.
   a) How much will it cost to use a 1 kW microwave if it takes 5 minutes and electricity is 13p/kWh?
   b) How much will it cost to use a 2 kW gas oven if it takes 30 minutes and gas is 4p/kWh?
   c) Comment on the difference in the costs.
7. Jo has just received her gas and electricity bill (Figure 1.3.3a) and has decided to find ways of reducing what she has to pay. Suggest three things she could consider doing to reduce her energy consumption.

FIGURE 1.3.3b: The cost of 1 kWh of energy from electricity is roughly three times that of 1 kWh of energy from gas.

Know this vocabulary

kilowatt-hour (kWh)

SEARCH: energy costs in the home

Energy

Getting the electricity we need

We are learning how to:
- Describe ways of generating electricity.
- Explain advantages and disadvantages of different methods.
- Evaluate the consequences of using different generating methods.

There are many different ways of generating electricity – anything that will make a magnet rotate inside a coil of wire will make a current flow. You could even use a hamster in a wheel, though you wouldn't get much of an output. We use vast amounts of electricity so supplying it cheaply and reliably is a big challenge.

Generating electricity

There are three main ways of generating electricity. The first is to burn a **fossil fuel**, such as coal, to heat water and turn it into steam. The steam drives a turbine (rather like a water wheel) which powers the generator. These are **non-renewable** energy sources. Fossil fuels are formed from animal and plant material over millions of years and contain a lot of energy, which is released upon combustion. The process releases polluting gases and supplies are running out.

The second is to use a nuclear reactor to heat water and produce steam, which then drives a turbine generator. Nuclear power is non-renewable as it uses fuel such as uranium, which has to be mined. Reactor waste remains dangerous for many years, and leaks can contaminate large areas of land.

The third way is a whole group of methods, referred to as **renewable** generation. Unlike fossil fuel and nuclear-powered generation, these methods don't use up fuels. They include:
- solar power, which produces electricity from sunlight;
- wind power, which uses the wind to turn blades and drive a generator;
- hydroelectric power, which uses falling water from a dam to drive a generator;
- wave farms, which extract energy from water waves;
- biomass, producing gas from decomposing animal or plant matter;
- geothermal, using heat from under the ground;
- tidal, extracting energy from the motion of water as the tide comes in and goes out.

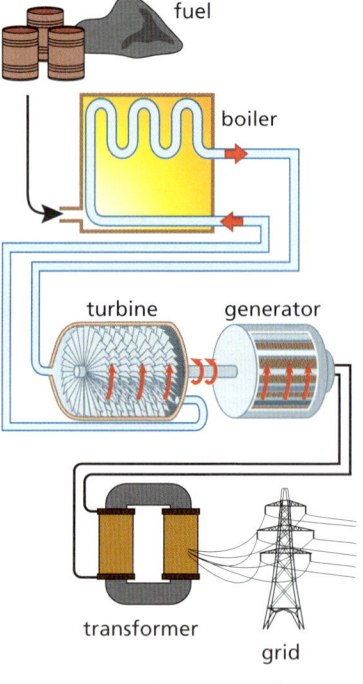

FIGURE 1.3.4a: The process of generating electricity.

Did you know...?

Renewable energy sources are not necessarily very efficient. Wind turbines, for example, only extract about 30% of the available energy from the wind. However, as the wind is free and there are no waste products this doesn't cause any problems. You could research other energy resources to see how they compare.

3.4

1. What other types of fossil fuel are there apart from coal?
2. What are the disadvantages of:
   a) Fossil fuel power stations?   b) Nuclear power stations?

Making decisions about which generating method to use

Governments have to decide how to produce the electricity that people need. They have to think about various factors such as:
- cost of construction;
- cost of operation;
- effect on the environment;
- availability of fuel (or sun, wind, etc.);
- reliability of the supply;
- safety of operation.

Sometimes there are difficult choices to be made. A hydroelectric power station, for example, needs no fuel and produces no waste materials but is expensive to build and affects a large area of land, flooding a valley and damaging habitats.

3. What problems are caused by burning lots of fossil fuels?
4. Quite a few people are not happy at the thought of using nuclear power stations. Why might this be?
5. The supply of power needs to be reliable. Why might solar power and wind power not be suitable as a sole energy source?

Using evidence to make a decision

Table 1.3.4 shows how much it costs to produce electricity by different methods in the UK. It shows how much it costs to produce 1 MWh (this is 1000 kWh; 1 kWh will run a small electric fire for an hour). As this shows, the costs vary quite a lot and they aren't fixed for one particular energy resource but vary over a range. Nevertheless, it's clear that some methods are cheaper than others.

Type of power station	Cost: £/MWh
Gas	80
Coal	102
Coal with carbon capture	122
Nuclear	81
Offshore wind farm	118–134
Solar	169
Onshore wind farm	93–104
Biomass	117–122

TABLE 1.3.4: Taken from Electricity Generation Costs, Dept of Energy & Climate Change, 2012, 'Table 1: Levelised cost estimates for projects starting in 2012, 10% discount.'

6. Renewable generating methods don't release harmful gases or make other waste materials. Why then do some people say they damage the environment?
7. Why not build the entire electricity-generating system using gas turbine stations?
8. Why might a gas turbine power station be built 'with CO_2 capture'?
9. Why do you think offshore wind power farms are more expensive to build and run than onshore ones?
10. Coal-fired power stations are around 45% efficient whereas wind turbines are around 38% and photovoltaic cells are about 21%. Explain whether this should persuade us to use more coal.

Know this vocabulary

fossil fuel
non-renewable
renewable

SEARCH: generating electricity   61

Energy

Using electricity responsibly

We are learning how to:
- Apply the concept of energy transfer to a wind-up device.
- Critique claims made for the running costs of fluorescent light bulbs.
- Evaluate actions that could be taken in response to rising energy demand.

It is easy to expect governments and energy suppliers to make good decisions about how we are supplied with energy. It is important that they do so, but as consumers we can also make decisions about how we use electricity. We have to get that right too.

Turning our work into light

The torch in Figure 1.3.5a has batteries in it but instead of buying them charged up or recharging them from the mains, you turn the crank. This drives a generator, which charges the batteries up. Then, when you want to use the torch, you switch it on like any other torch. When the bulb starts to go a bit dim you can recharge it with the handle.

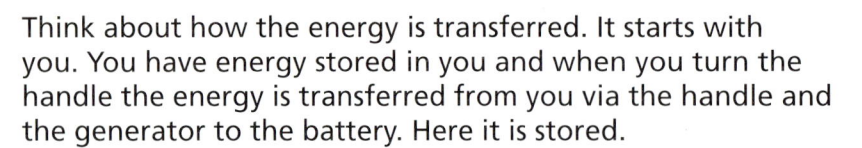

FIGURE 1.3.5a: A wind-up torch.

Think about how the energy is transferred. It starts with you. You have energy stored in you and when you turn the handle the energy is transferred from you via the handle and the generator to the battery. Here it is stored.

Then when you switch the light on the energy is transferred from the battery, via the wires and the bulb to the environment. It's now spread out, or **dissipated**, widely. The energy is all still there but it's of little use.

1. What would happen if you turned the handle for longer?
2. What difference would it make if a torch had two bulbs that lit up instead of one?
3. Is it true to say that by using this device you are 'generating energy'?

Investigating claims for low-energy light bulbs

Low-energy light bulbs have been around for a few years now but when they first came out manufacturers had to promote them. They were more expensive than the filament bulbs they replaced. Adverts like the one in Figure 1.3.5b are very appealing – they suggest that what you save in lower electricity bills more than pays for the cost of the more expensive bulb. Is it true though? In order to compare the bulbs, we need to allow for a number of factors, such as:

- how much the bulb costs;
- how much energy the bulb uses;
- how much it's going to be used;
- how long the bulb lasts for;
- how much electricity costs.

3.5

Assume we can buy a fluorescent bulb for £5 and a filament bulb for £1. The fluorescent bulb is rated at 20W and the filament bulb at 100W, but they both give out a similar amount of light. The fluorescent bulb will be good for 10 000 hours and the filament bulb for 1000 hours. Electricity costs 13p/unit and the bulb will be used for around 3 hours a day.

Let's see what they cost to run for a year, which will be the approximate life of the filament bulb. The filament bulb costs £1 and will use 0.3 units of electricity per day, which will cost 3.9p. Over the year this will total £14.24, plus the cost of the bulb, so £15.24.

The fluorescent bulb will use 0.06 units a day, which will cost 0.78p. Over one year this comes to £2.84. We need to add in the cost of the bulb, but remember that this kind of bulb will last for 10 000 hours. In the year we've only used it for about 1000 hours so we should only count one-tenth of its cost, which is 50p. The total is therefore £3.34.

FIGURE 1.3.5b: Adverts similar to this encouraged people to buy low-energy light bulbs.

4. What conclusion can you draw from the data?
5. Is the manufacturer's claim true?
6. Explain why the fluorescent bulb is cheaper to use, even though it's much more expensive to buy?
7. Instead of a fluorescent bulb you could use an LED bulb, which costs twice as much but lasts for twice as long and uses half the amount of electricity. Do the calculations for the use of the LED bulb over a year, to see how it compares.

Deciding on actions to take

Demand for electricity rises as population increases and people use more appliances. How is this dealt with? Sometimes actions are taken at national level and sometimes at local level. There are two areas to consider:

- How energy is supplied to a community. Is it being generated in a way that is not only cheap and reliable but doesn't damage the environment?
- How energy is used in the community. Are users being economical with their use of electricity, selecting efficient appliances and using them responsibly?

8. What kind of electricity generation system do you think we should be using? Justify your answer.
9. How can consumers be encouraged to use electricity more responsibly?
10. Efficient appliances, such as fridges with better insulation, sometimes cost more. How could you persuade people that they were a good idea?

Know this vocabulary

dissipated

SEARCH: using electricity efficiently 63

Energy

Energy stores and transfers

We are learning how to:
- Use a model of energy.
- Describe energy stores and transfers.
- Apply the energy model to different situations.

Energy is a really important concept in science and is used in many different areas. Explaining everything from food chains and chemical reactions through to roller coasters and electric circuits involves using ideas about energy. It's important to understand that although energy is used to explain many different effects, it's all just energy – there aren't different types of energy.

A model of energy

A **model** is used in science to help us to understand a concept. A model for energy that is used is the 'stores and transfers' model. This helps us to make sense of what energy does. The model uses the idea that energy is in a 'store'. The energy may be transferred and then it will end up in another store. We can think of these stores being emptied as energy is transferred out of them, and filled as energy is transferred into them.

For example, if a teacher is lifting boxes of books onto a shelf, energy is being transferred. The teacher is a **chemical energy store** and as she does work the energy is transferred from that store, via the reactions that make muscles work, to the **gravitational potential energy store** in the boxes as they are raised up. If one of the boxes falls off the shelf, energy is transferred out of the gravitational potential energy store. The energy will then be **dissipated** to the surrounding environment – it will be spread out wastefully. The environment is a store too.

FIGURE 1.3.6a: This boy is using energy to lift these books – but what type of energy?

1. How would the amount of energy transferred differ if the boxes are put on a higher shelf?
2. How would the amount of energy transferred differ if more boxes are moved?

Examples of stores and transfers

- Using a cooker to heat food. Energy is transferred from a chemical energy store (fuel and oxygen) to a **thermal energy store** in the food. The more energy that

Did you know…?

Every object that is raised up is a store of energy, because things can happen when it falls! This is because of gravity, and we call this store a gravitational potential energy store.

64 AQA KS3 Science Student Book Part 1: Energy – Energy costs *and* Energy transfer

is transferred, the higher the temperature attained by the food.

- A child on a swing. Energy is transferred from the gravitational potential energy store (at the top of the swing) to a **kinetic energy store** of movement and then back to the gravitational potential energy store (at the other end). Swinging back and forth repeatedly involves many transfers and the stores filling up and emptying repeatedly.
- A person on a trampoline. Energy is transferred from the kinetic energy store as the person moves downward and is slowed down by landing and into an **elastic energy store** as the trampoline springs are stretched. Almost immediately the energy is transferred back again into a kinetic energy store.

3. A box falling off the shelf adds to the thermal energy store of the environment. Why is it difficult to measure this change?

4. Suggest another example of a device that stores energy in an elastic energy store.

5. Not all of the energy from the cooker will be transferred to the food; where will some of it be transferred to?

6. The kinetic energy store of the child on the swing keeps dropping to zero, at each end of the swing when they are momentarily stationary. But their gravitational potential energy store is never zero, even at the mid-point of the swing. Explain why.

FIGURE 1.3.6b: Examples of stores and transfers.

Applying the model

The 'stores and transfers' model can be used to describe what is happening in a variety of situations. See if you can apply it to these:

7. A pendulum swings back and forth.
8. A ball is dropped and bounces back up again.
9. The pendulum eventually stops swinging.
10. The bouncing ball eventually comes to rest on the ground.
11. Alex says 'There's water in this jug and I can pour it into a cup. All the water is still there but it's been transferred to a different store.' How good is this way of explaining what happens to energy when it is transferred between stores?

Know this vocabulary

model
chemical energy store
gravitational potential energy store
dissipated
thermal energy store
kinetic energy store
elastic energy store

SEARCH: energy stores and transfers   65

Energy

Exploring energy transfers

We are learning how to:
- Recognise what energy is and its unit.
- Describe a range of energy transfers using simple diagrams.
- Use a Sankey diagram as a model to represent simple energy changes.

The Sun is our main source of energy. Plants convert this energy by chemical processes to make food. Solar panels transfer the Sun's energy by electric current to provide electricity for our use. By transferring energy from the Sun, useful energy can be provided for our planet.

Thinking about transfers

When **energy** is transferred, useful things can happen. When a log is burned, energy is transferred by chemical reactions to the surroundings by light and heat. Switching on a light bulb transfers energy by electric current to the bulb. Energy is then transferred from the bulb to the surroundings increasing the temperature.

Energy is never lost or made; it is just transferred to the surroundings, increasing the temperature.

1. Look at the photos on this page. In which of them is energy being transferred?

2. a) What is happening as a result of the energy transfer you can see in Figure 1.3.7b?

   b) What is happening in the other photo in Figure 1.3.7b? Why is it not possible for energy to be transferred here?

FIGURE 1.3.7a: Where has the energy to light this bulb come from?

Energy transfers

It is useful to track the processes by which the energy is transferred. This can be done using a simple **energy transfer diagram** (see Figure 1.3.7c). When you switch on a light bulb, you want to transfer energy by light. However, the light bulb also gets hot. Transferring energy by heating is not useful in this instance. Energy-**efficient** light bulbs have been designed to transfer more energy by light and less by heating.

FIGURE 1.3.7b: Describe the differences, in terms of energy transfer.

energy stored in the generating system → electricity in the wires → energy stored in the surroundings

FIGURE 1.3.7c: Simple energy transfer diagram for a light bulb.

3. Write a sentence to describe the energy transfers shown in Figure 1.3.7c.

4. Draw a diagram to show how energy is transferred by:
   a) a boiling kettle;
   b) a toaster;
   c) a log fire.

5. In your answers to question 4, underline the useful energy transfers and circle the unwanted energy transfers.

Did you know…?

The amount of energy transferred to the Earth from the Sun in one minute is enough to meet the world's energy demands for one year.

Sankey diagrams

If you move a weight of 1 N through a distance of 1 m, you transfer 1 joule (1 J) of energy. One joule of energy is also needed to heat 1 cm³ of water by 1 °C.

A **Sankey diagram** is a type of energy transfer diagram that shows the relative amounts of energy transferred by a device. The width of each arrow shows how much energy is transferred. The non-useful energy transferred is always shown pointing downwards. The greater the proportion of energy transferred that is useful, the more efficient we say the device is.

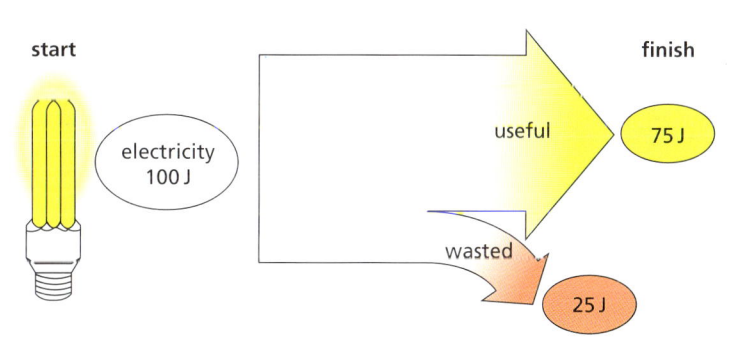

FIGURE 1.3.7d: Sankey diagram for an energy-efficient light bulb. How would the Sankey diagram for an old-style, less efficient light bulb compare with this one?

For example, in Figure 1.3.7d, 100 J of energy is transferred to the light bulb by electric current. It transfers 75 J by light (useful) and 25 J by heating the surroundings (wasted). If you draw these on graph paper, you can accurately represent the proportions of energy involved.

6. What is the percentage of energy wasted in the light bulb in Fig 1.3.7d?

7. On graph paper, draw a Sankey diagram for an electric drill that transfers 500 J of energy. 300 J of energy is transferred as movement and 200 J are transferred to the environment. You will need to decide which of the outputs are useful and which are useless.

8. What would make the drill in question 7 more efficient?

Know this vocabulary

energy transfer diagram
efficient
Sankey diagram

Energy

Understanding potential energy and kinetic energy

We are learning how to:
- Recognise energy transfers due to falling objects.
- Describe factors affecting energy transfers related to falling objects.
- Explain how energy is conserved when an object falls.

Many theme parks make use of energy transfer in their rides. An object high up has the potential to transfer energy. There is a new vertical-drop ride, the 'Drop of Doom', which, at 126 metres tall, is the tallest ever. People will fall from a stationary position at the top and reach speeds of up to 150 km per hour.

FIGURE 1.3.8a: *The Zumanjaro: Drop of Doom*, theme park ride. How is energy being transferred as people drop from the top to the bottom of this ride?

What is gravitational potential energy?

Objects at a height possess energy, because of the Earth's gravitational field – think of parachute jumpers or sky divers. Their **gravitational potential energy store** is pretty full. This energy is transferred when the object loses height.

1. What is the unit for gravitational potential energy?
2. What is the name of the force acting on objects that causes them to have gravitational potential energy?

Factors affecting gravitational potential energy

The higher an object is, the more gravitational potential energy it has. More energy can be transferred to make it move. When the objects falls, energy is transferred from the gravitational potential energy store to the **kinetic energy store**. As it does so, the object moves faster and faster.

It is useful to think about objects acting as energy stores, which can be filled up in different ways. For example, the kinetic energy store is filled as an object speeds up and the gravitational potential energy is filled when an object is raised.

The greater the force acting on the object, the more energy that can be transferred. The force of gravity is greater on Jupiter than on Earth, so an object falling the same distance on Jupiter will transfer more gravitational potential energy than it would on Earth.

FIGURE 1.3.8b: How does gravitational potential energy affect these people?

68   AQA KS3 Science Student Book Part 1: Energy – Energy costs *and* Energy transfer

3. A tennis ball falls from the following heights:
   i) 10 mm   ii) 10 cm   iii) 10 m
   a) Represent this by an energy transfer diagram showing stores and transfers.
   b) Which fall will transfer the most energy?

4. Look at Table 1.3.8. If a tennis ball is dropped from the same height on each planet, on which planet will it reach the highest speed?

TABLE 1.3.8: Gravitational field strengths on different planets.

Planet	Gravitational field strength (N/kg)
Earth	10
Mars	3.7
Saturn	11

Conservation of energy in falling objects

Gravitational potential energy is transferred by movement and heating the surroundings. As a falling object drops lower, its gravitational potential energy decreases and the amount of energy transferred to kinetic energy increases. Some energy will also be transferred by heating the surroundings, due to friction with the air particles during the fall. The faster the object falls, the greater the energy transferred by heating. When the object hits the ground, some of the kinetic energy may stay in it if it bounces back up but the rest is transferred by heating and sound to the surroundings.

5. Look at Figure 1.3.8c of a ball falling from a height. In which position (A, B or C) does the ball have:
   a) The greatest amount of energy in the gravitational potential energy store?
   b) The least amount of energy in the gravitational potential energy store?
   c) The least amount of energy in the kinetic energy store?
   d) The greatest amount of energy in the kinetic energy store?

6. Sketch two graphs to show how the energy levels in the gravitational potential energy store and the kinetic energy store of the ball in Fig 1.3.8c change during the fall.

Did you know...?

The Stealth roller coaster at Thorpe Park has the greatest acceleration of any such ride in the UK. Riders accelerate from rest to 130 km/h in under 2 seconds, propelling them to a height of 62.5 m. Occasionally energy losses mean the train fails to reach the peak and (safely) rolls back.

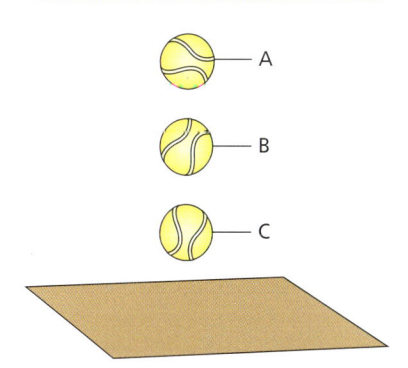

FIGURE 1.3.8c: A ball transferring gravitational potential energy.

Know this vocabulary

gravitational potential energy store
kinetic energy store

SEARCH: gravitational potential energy

Energy

Understanding elastic energy

We are learning how to:

- Describe different situations that use the energy stored in stretching and compressing elastic materials.
- Describe how elastic energy in different materials can be compared.
- Explain how elastic energy is transferred.

Elastic materials have the ability to store energy ready for use. The muscle tissue in animals consists of fibres of protein that can expand and contract, providing a potential store of elastic energy. This ability allows us to jump and move – and allows fleas to jump more than a hundred times their own height!

FIGURE 1.3.9a: A flea's jump is an example of elastic energy being transferred.

What is elastic energy?

Energy is stored when an elastic material is stretched or compressed (squashed) by a force. You do work when you pull an elastic band or squash a spring. This transfers energy, which is stored in an **elastic energy store**.

The stored energy is transferred when the elastic material returns to its original shape.

The further a material is stretched or compressed, and still be able to return to its original position, the more energy can be transferred.

1. In which of the situations in Figure 1.3.9b is more elastic energy transferred?
2. What causes the jack-in-the-box to bounce up when the lid is opened?

Applications of elastic energy

Catapults and archery bows use elastic materials. Elastic energy is stored when the elastic is stretched or the bow is bent. More elastic energy is stored if the elastic is harder to stretch because more work is done in pulling it back.

Some shock absorbers in cars have strong springs. When driving over a bump, energy is transferred by movement into the elastic energy store in the springs. This energy is released slowly when the car gets beyond the bump.

FIGURE 1.3.9b: What do these have in common?

70   AQA KS3 Science Student Book Part 1: Energy – Energy costs *and* Energy transfer

3. Some students are testing two different elastic materials for use in a catapult. They want to find out which would transfer more energy.
   a) How should they make the investigation a fair test?
   b) What should they measure to collect evidence?
4. Describe the energy transfers in a wind-up clock and represent this on an energy transfer diagram showing stores and transfers.

Did you know…?

Many elastic materials can stretch up to five times their original length. The first type of elastic material was natural rubber, made from the sap of rubber trees. Scientists have recently invented a gel material that can stretch up to 20 times its original length and still recover. It has a possible application as artificial cartilage, because it is also extremely strong.

Explaining elastic energy

Elastic materials, such as rubber, are made up of **molecules** that are bound together. When the material is stretched, the bonds between the molecules store energy.

In its relaxed state, rubber consists of long strands of molecules which are all coiled up. When the rubber is stretched, the coils become elongated and straightened, enabling the rubber to extend in length. When the stretching force is removed, the molecules return to their coiled-up state and the material returns to its original length.

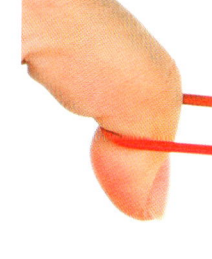

FIGURE 1.3.9c: A rubber band in a relaxed and stretched state.

The elastic energy stored in a rubber band or a spring is equal to the energy transferred in stretching it. This energy can be transferred as kinetic energy when the stretching force is removed.

5. Can all materials store elastic energy? Explain your answer.
6. How would you test which had more elastic energy – a coiled metal spring or an elastic band?

Know this vocabulary

elastic energy store
molecules

SEARCH: elastic energy

Energy

Checking your progress

To make good progress in understanding science you need to focus on these ideas and skills.

- ☐ Describe how jobs get done, using an energy model where energy is transferred from one store to another.

- ☐ Explain that energy is transferred from one type of energy store to another when change happens.

- ☐ Explain that all changes, physical or chemical, result in a transfer of energy.

- ☐ Recall that energy is measured in joules.

- ☐ Explain that it is sometimes better to measure energy in kilojoules or kilowatt hours.

- ☐ Carry out calculations of quantities of stored and transferred energy.

- ☐ Describe what is meant by rate of energy transfer.

- ☐ Identify the rate at which electrical appliances transfer energy (their power rating), using the correct units (watts or kilowatts).

- ☐ Compare rates of energy transferred when electrical appliances are used.

- ☐ Use the power rating of an appliance to calculate the amount of energy transferred.

- ☐ Compare the energy usage of different appliances.

- ☐ Calculate the cost of energy usage: cost = power (kW) × time (hours) × cost (pence per kWh).

- ☐ Recognise that energy is transferred by a range of different processes.

- ☐ Interpret and draw energy transfer diagrams for a range of different energy transfers.

- ☐ Use Sankey diagrams to explain a range of energy changes and demonstrate that all energy is always accounted for.

3.10

- ☐ Identify simple energy transfers that involve gravitational potential, elastic, kinetic, thermal and chemical energy.
- ☐ Explain how energy is transferred using elastic, chemical and gravitational potential energy.
- ☐ Analyse changes in gravitational potential energy in different situations.

- ☐ Recognise that electricity is generated in a variety of ways.
- ☐ Describe advantages and disadvantages of various ways of generating electricity.
- ☐ Use data to evaluate social, economic and environmental consequences of a particular way of generating electricity.

- ☐ Give examples of renewable and non-renewable energy resources.
- ☐ Explain the advantages and disadvantages of renewable and non-renewable energy resources.
- ☐ Explain the challenges involved in moving towards a more renewable energy supply system.

- ☐ Identify how appliances that transfer energy result in some energy being dissipated, reducing the useful energy.
- ☐ Suggest ways in which energy dissipation in a process could be reduced.
- ☐ Suggest ways in which a home energy bill could be reduced.

- ☐ Understand that food is a fuel.
- ☐ Explain that food labels provide information about the different amounts of energy in various foods.
- ☐ Explain that energy is transferred from the chemical energy store when we perform physical activities.

Energy

Questions

KNOW. Questions 1–4

See how well you have understood the ideas in this chapter.

1. Which of the following is a unit of energy? [1]
   a) kilogram   b) kilojoule   c) kilometre   d) kilohertz

2. Which of the following is *not* a fuel? [1]
   a) petrol   b) sugar   c) coal   d) air

3. State two ways that energy can be stored. [2]

4. Describe the energy transfer when a ball falls from a height. [2]

APPLY. Questions 5–7

See how well you can apply the ideas in this chapter to new situations.

5. Describe the energy transfers that happen when an archer pulls back and then releases a bow to shoot an arrow. [2]

6. Describe the energy transfers that occur as a burning gas drives an electricity generator. [2]

7. Nutritional information about food products is shown on their labels, including the energy stored. Table 1.3.11 shows some information about different types of milk. What does this tell you about the differences between whole, semi-skimmed and skimmed milk? [2]

TABLE 1.3.11

	Unit	Amounts in 100 cm³ of milk		
		Whole (full cream)	Semi-skimmed	Skimmed
Energy stored	kilojoule (kJ)	282	201	148
Protein	gram (g)	3.4	3.6	3.6
Carbohydrate	gram (g)	4.7	4.8	4.9
Fat	gram (g)	4.0	1.8	0.3

EXTEND. Questions 8–10

See how well you can understand and explain new ideas and evidence.

8. Julia's science teacher tells her that 'energy-efficient' light bulbs are better to use because they waste less energy through heating the surroundings. But Julia knows that her mother, who is a farmer, uses old-fashioned filament light bulbs to keep newly hatched chicks warm in winter. Which of these statements is correct? [1]

    a) Julia's teacher is right – bulbs that transfer most of the energy by heating the surroundings are always wasteful.
    b) The chicks don't need heating – they just need to see where they are going.
    c) Heating the surroundings is only wasteful if you don't make use of it.
    d) Julia's mother should switch to energy-efficient light bulbs.

9. Explain why an electric kettle has a power rating of 2000 W, but a small TV has a power rating of 65 W. [1]

10. You are looking for the best possible fuel source for the future. Use the data in the table to make your choice. Give reasons for your answer. [2]

Fuel	Energy per gram (J/g)	State	Harmful products of combustion	Availability
coal	24	solid	carbon dioxide, soot, acid rain	running out
hydrogen	123	gas	none	plenty
petrol	46	liquid	carbon dioxide	running out
biofuel	33	liquid	carbon dioxide	renewable

Waves
Sound *and* Light

Ideas you have met before

Different types of sound

Sounds are only possible when a vibration occurs. Banging on a drum or plucking a guitar produces vibrations that cause a sound to be made.

We can change the vibrations of a sound by giving them more energy. The stronger the vibrations, the louder the sound.

Some sounds we hear have a high pitch, like a whistle or a siren. Some have a low pitch, like the rumble of thunder. When we change the pitch, we change how rapidly an object vibrates.

How sounds behave

We hear sounds because the vibrations travel through a material, like air, to the ear.

Sounds may be reflected by hard materials and absorbed by soft materials.

Sounds get fainter as they travel further from the source.

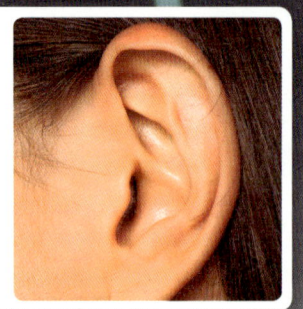

How light behaves

Light appears to travel in straight lines.

Shadows have the same shape as the objects that made them because of light travelling in straight lines.

How we see things

We see objects because they emit or reflect light into our eyes.

We can see objects that don't emit their own light because they reflect light from other sources into our eyes.

We can explain this using the idea that light travels in straight lines.

AQA KS3 Science Student Book Part 1: Waves – Sound *and* Light

4.0 In this chapter you will find out

What sound is
- Energy is transferred by sound in the form of waves.
- Sound travels as longitudinal waves (vibrations) passed on by particles of a material.
- Sounds can be represented by waveforms, showing wavelength, frequency and amplitude.
- The greater the amplitude of the waveform, the louder the sound.
- The greater the frequency (and the shorter the wavelength), the higher the pitch.
- The ear is a detector of sound waves of a certain frequency range.

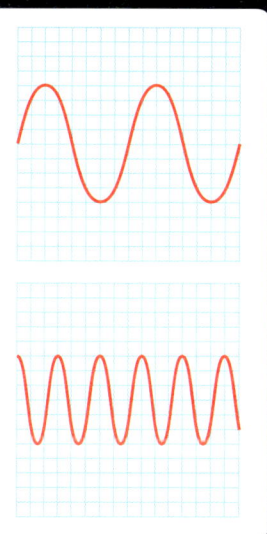

How sound behaves
- The denser the medium, the faster sound travels.
- Sound is transmitted, reflected or absorbed by different types of surface.
- Echoes occur when sound waves are reflected by hard materials.

What light is
- Light travels as transverse waves that carry energy.
- White light can be split into a spectrum of colours.
- Coloured light causes an object to appear a different colour.

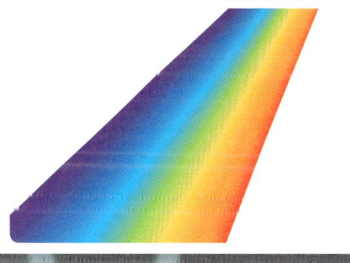

How light behaves
- Light waves can travel through a vacuum.
- Light can be reflected, absorbed and refracted.
- When it is reflected, the angle of incidence equals the angle of reflection. Light can form an image in a mirror.
- Light can be refracted through lenses and prisms.
- Wave properties can be described using a ray diagram as a model.

77

Waves

Exploring sound

We are learning how to:
- Identify how sounds are made.
- Describe how sound waves transfer energy.
- Explain how loud and quiet sounds are made.

Sounds are made in different ways and by many different things. We need to understand what sound is, what all sounds have in common and how they vary.

Making sounds

If you place a finger over your voice box when speaking or singing, you will feel the **vibration** of your voice box. This is where the sound comes from.

When an instrument is plucked or blown through, the string or the air vibrates. Often the vibrations are too small to see.

All vibrations result in a sound. The vibrations from the object are passed on to air particles. These air particles bump into others and the wave progresses. Eventually the energy of the vibrations is transferred to your ears. The speed of sound in air is just over 343 m/s, around a million times slower than light.

1. What causes the sound when a bell is rung?
2. How does the sound from a concert reach the audience?

FIGURE 1.4.1a: How does a guitar make a sound?

Making waves

Energy is transferred by sound in the form of waves. In Figure 1.4.1b a slinky spring provides a model that shows how these waves work. When you push the end of a slinky back and forth, some of the coils squash together and others pull apart. A wave of energy passes along the length of the spring. A wave like this which travels in the same line as the vibrations of the source is called a **longitudinal wave**.

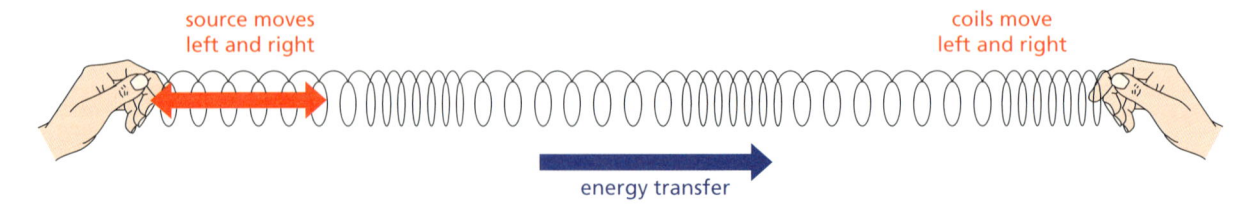

source moves left and right

coils move left and right

energy transfer

FIGURE 1.4.1b: Why is this called a longitudinal wave?

A sound wave works in the same way as the slinky spring. Vibrations push air particles together and also pull them apart, creating a longitudinal wave of energy. The energy is transferred from the source of the vibration to our ears.

3. Describe the movement of air particles in a longitudinal sound wave.
4. Explain how a longitudinal wave can transfer energy from one store to another.

Loudness of sounds

The **volume** of a sound a measure of how loud the sound is. Sounds can be made louder by increasing the energy in the vibration. Plucking a string harder, blowing harder through a wind instrument or beating a drum harder will all transfer more energy. The loudness of sound is measured in a unit called a **decibel (dB)**. The loudest sound that humans can listen to without damage to their hearing is about 120 dB.

The size of a vibration is represented by its **amplitude**. Figure 1.4.1d shows that the amplitude is the maximum distance that a particle travels, from its middle position, in the to-and-fro vibration. The greater the amplitude, the greater the energy of the vibration and the louder the sound. In other words, a bigger wave will transfer more energy and be heard as a louder sound.

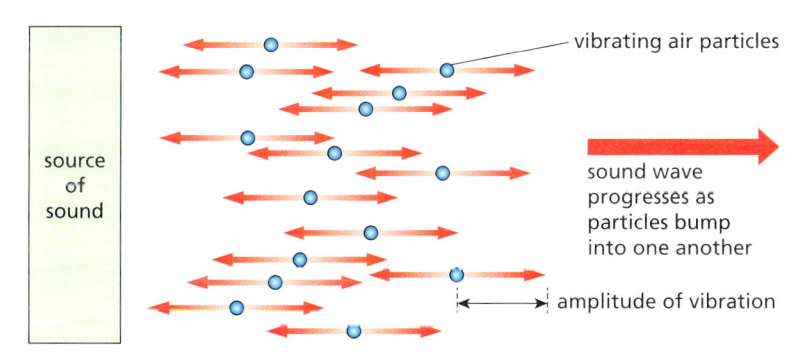

FIGURE 1.4.1d: What effect will a smaller amplitude have?

5. Look at Table 1.4.1. Match the sounds to the correct loudness.
6. The loudness of a sound also depends on the distance from the source. Explain what happens to the energy as you get further away.

4.1

Did you know...?

The ocean-dwelling tiger pistol shrimp is known to produce sounds of over 200 dB. It uses the sound as a defence mechanism. The vibrations can kill prey and fish up to 2 metres away!

FIGURE 1.4.1c: A pistol shrimp.

TABLE 1.4.1

Sound	Loudness (dB)
1 whisper	a) 80
2 phone ringtone	b) 140
3 jet engine	c) 100
4 motorbike	d) 30

Know this vocabulary

vibration
longitudinal wave
volume
decibel (dB)
amplitude

SEARCH: sound waves, longitudinal waves   79

Waves

Describing sound

We are learning how to:
- Explain what is meant by pitch.
- Understand frequency, wavelength and amplitude.
- Relate sounds to displayed waveforms.

There are many different types of sound. Think of the sounds made by a whale compared with the high-pitched screeching of a monkey, or the sound of a bass guitar compared with a violin. Differences in sound waves arise from different characteristics of the sound waves.

What is pitch?

A ship's horn produces a sound that is very deep and low – this is known as a low **pitch**. Whistles, alarms and sirens produce high-pitched sound.

The pitch of a note depends on the **frequency** of the vibration producing the sound. A high frequency means more vibrations are produced per second. Frequency is measured in the unit **hertz**, abbreviation Hz; 1 Hz is one vibration per second. A high frequency gives a high pitch, and a low frequency gives a low pitch. Musical notes change in pitch by changing the frequency of the vibration. Feel your voice box as you make sounds of different pitches. What do you notice?

1. Describe one other sound with a low pitch and one other sound with a high pitch.
2. What is meant by the 'frequency' of a note?

Wavelength, frequency and amplitude

Waves transfer energy. If no energy is being supplied, the wave can be represented by a horizontal straight line, like still water in a pond. If energy is then supplied, the wave starts to rise and fall – we say it is being displaced.

We can show this on a graph, such as Fig 1.4.2c. The amplitude shows how much the wave is being displaced vertically and the energy is being transferred from left to right.

The higher the frequency of a wave, the shorter the **wavelength**. The maximum displacement is the **amplitude**. The energy transferred by the wave depends on this. The larger the amplitude of a sound wave, the louder the sound.

FIGURE 1.4.2a: How would you describe the scream of a monkey?

FIGURE 1.4.2b: Energy is being transferred by waves; the water is being displaced above and below its level when at rest.

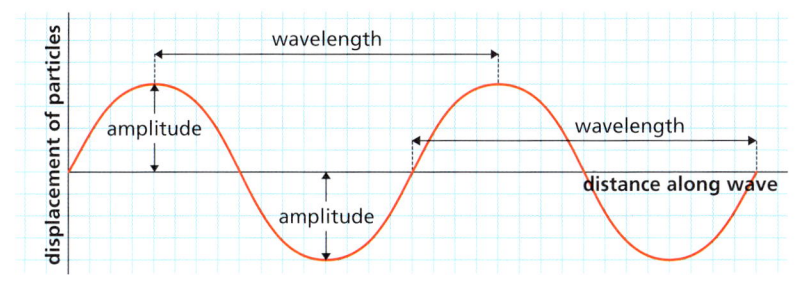

FIGURE 1.4.2c: Parts of a wave.

3. Why is it more useful to use the wave representation in Figure 1.4.2c, compared with a drawing of a longitudinal vibration, as in Figure 1.4.1d?

4. How could you tell from a **waveform** whether a sound is getting:
   a) louder?   b) higher pitched?

Interpreting sound waves

All sound waves can be detected using a microphone and shown as a waveform on a screen of a device called a cathode ray **oscilloscope** (or CRO). The microphone receives the sound waves and converts them into electrical signals. Some typical examples are shown in Figure 1.4.2d. The waveforms produced by a CRO are very useful as they show both the amplitude and the frequency of a sound wave. As the sound alters, so the waveform displays changes.

5. a) Which wave in Figure 1.4.2d results from the loudest sound?
   b) Which wave results from highest-pitch sound?
   c) Which wave is transferring the most energy? Explain your answer.

6. Draw waves to represent a loud high-pitched flute note and a quiet low-pitched flute note.

7. Look at the graph in Figure 1.4.2e, which shows the sound wave detected from a gun as time progresses. Describe what is happening to the frequency, wavelength and amplitude of the wave.

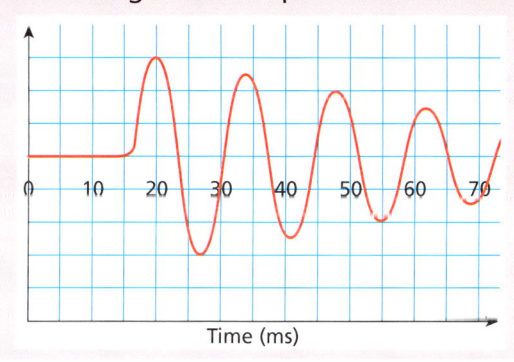

FIGURE 1.4.2e: Sound wave from a gun

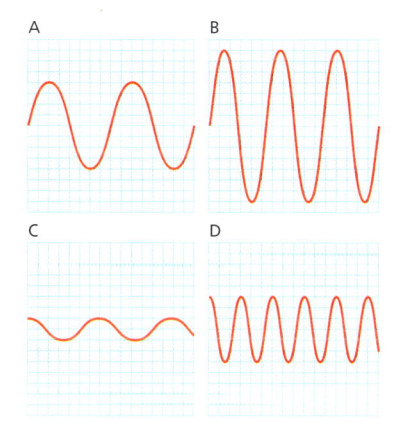

FIGURE 1.4.2d: How are these waves different?

Did you know…?

Microphones have a thin diaphragm made of plastic or metal. This vibrates when even small sound vibrations hit it. These vibrations are converted into an electrical signal that can be fed to a loudspeaker. The electrical signal is then converted back into vibrations, which are heard as sound.

Know this vocabulary

pitch
frequency
hertz (Hz)
wavelength
amplitude
waveform
oscilloscope

SEARCH: wavelength, frequency and amplitude of sound waves   81

Waves

Hearing sounds

We are learning how to:
- Explain what is meant by audible range.
- Understand how the ear detects sounds.
- Apply ideas about sound to explaining defects in hearing.

The ability to hear is important in all animals for communication, hearing predators, knowing when there is danger and seeking prey. The human ear relies on a combination of processes and perfectly evolved mechanisms to allow us to identify the wide range of sound waves we receive.

What can we hear?

We know that some notes are higher than others and what those high notes are like. Some notes can be painful to listen to, such as very high ones. In fact, there's a limit as to what we can hear. Sounds are made by things vibrating and the number of times per second something vibrates is the frequency. Humans can usually hear frequencies up to around 20 000 Hz (20 kHz).

We can also hear frequencies down to 20 Hz so the human audible range (or **auditory range**) is 20 Hz to 20 kHz. Some animals can hear much higher frequencies – dogs up to 45 kHz and cats up to 64 kHz.

FIGURE 1.4.3a: Good hearing is essential for detection of predators.

1. How many Hertz in 1 kHz?
2. Why would you not be impressed by an advert for a personal stereo that claimed it had maximum frequencies of 30 kHz?
3. Dog whistles can be heard by dogs but not by humans. They can have a range of frequencies – but between what limits?

Structure of the human ear

The function (job) of the ear is to transfer energy by sound into electrical impulses that are interpreted by the brain. Figure 1.4.3b describes what happens to the sound waves as they enter the ear.

4. Suggest why incoming sound vibrations need to be amplified (amplitude made bigger) in the ear.
5. Where in the ear are:
   a) Electrical signals transmitted to the brain?
   b) Sound vibrations amplified?
   c) Vibrations first detected?

Did you know…?

Elephants can hear frequencies 20 times lower than the lowest frequency we can hear. They use their trunks as well as their ears to detect low frequency vibrations. This enables them to hear other elephants up to 6 km away.

4.3

Start here:
A vibrating object such as a trumpet produces sound waves.

The sound waves travel through the air to the ear.

The vibrating air enters the ear and is funnelled into the ear canal.

The vibrating air sets the ear drum vibrating.

The ear drum transmits the vibrations to the ossicle bones, which start to vibrate.

The vibrations of the ossicles are amplified and transmitted to the fluid in the cochlea.

Cells lining the cochlea detect the vibrations and convert them to electrical signals.

The auditory nerve transmits the electrical signals to the brain.

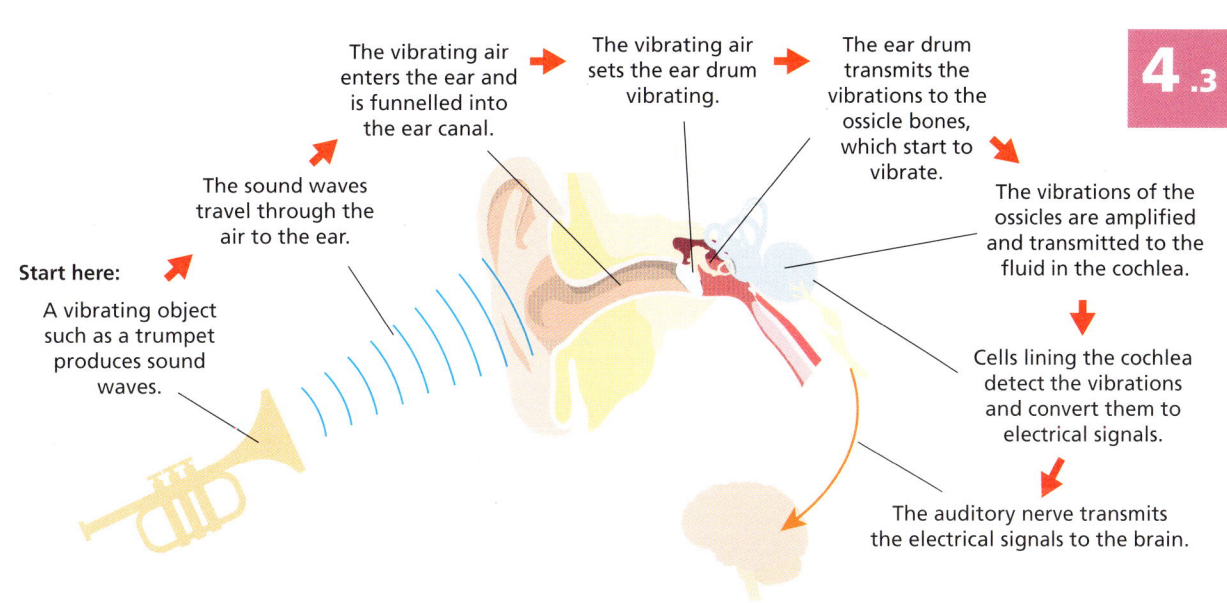

FIGURE 1.4.3b: How we hear.

Factors affecting hearing

Several factors can affect the health of our ears. Read about these in Table 1.4.3.

TABLE 1.4.3: Causes of ear damage and what can be done.

Causes of poor hearing or ear damage	Possible solutions
Ear canal can become blocked with wax.	Have the ear canal cleaned out.
Very loud sounds can rupture the ear drum.	Ear drum may heal itself over a long period of time.
Ear drum can be damaged by infection.	Use antibiotics to get rid of the infection.
Ossicles can become fused together.	An operation is needed.
Infection may occur in the middle ear.	Use antibiotics to get rid of the infection.
Hair cells and nerves in the cochlea may be damaged by loud noises.	There is no cure.
In older people, nerve cells may deteriorate.	There is no cure.

6. Why can some ear problems not be cured?
7. Who is most likely to be most at risk of having problems with poor hearing?

Know this vocabulary

auditory range

SEARCH: how the ear works

Waves

Understanding how sound travels through materials

We are learning how to:
- Recognise how the speed of sound changes in different substances.
- Explain why the speed of sound varies between solids, liquids and gases.

Whales are known to transmit sounds in the ocean over distances of 700 km. If whales were to transmit these same sounds in the air, would they travel faster or slower?

Sound in a vacuum

Most of the sounds that you hear are transmitted by vibrating air **particles** (particles of gas). Sounds can also travel through solids and liquids. Sound waves need particles of matter to transmit energy. As the particles vibrate, the energy is passed on to adjacent particles and carried in the form of a wave.

Sounds cannot travel through a **vacuum**, nor through space, which has hardly any particles in it.

FIGURE 1.4.4a: How do sounds from whales travel under water?

1. Why can sound not travel through a vacuum?
2. How is it possible for sounds to travel through solids?

Speed of sound through different materials

Table 1.4.4 shows the speeds that sound travels through different materials.

3. a) In which material does sound travel the fastest?
   b) In which material does sound travel the slowest?
   c) Does sound travel fastest in solids, liquids or gases?

TABLE 1.4.4: Speed of sound in different materials.

Material	Speed of sound (m/s)
air	343
carbon dioxide	259
copper	5010
diamond	12 000
lead	1960
oxygen	316
steel	5960
water	1482

84 AQA KS3 Science Student Book Part 1: Waves – Sound *and* Light

Sound and particles

4.4

Particles of matter in solids, liquids and gases differ in their arrangement and behaviour. This affects how well sound waves can travel through them. The speed at which the wave moves depends on the arrangement of the particles, the elastic nature of the forces between them, and how fast the particles are moving.

- In a gas the particles are very far apart. Sound travels slowly because the particles do not collide very often.

- In a liquid the particles are much closer to one another. Sound travels more quickly because the particles are able to collide with each other much more frequently. Sound travels about five times faster through liquids than it does through gases.

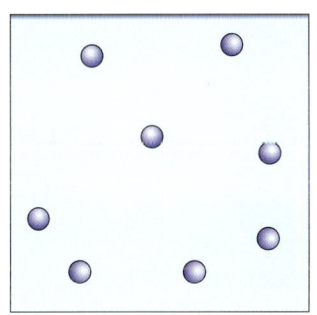

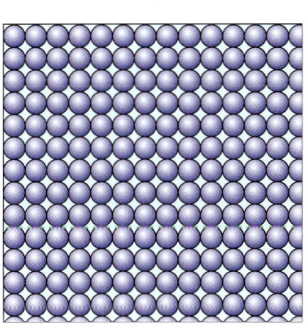

> **Did you know...?**
>
> Native Americans used to put an ear to railway tracks to know when trains were coming. This is a dangerous thing to do because you can never tell how soon the train will arrive.

Gas | Liquid
Solid | Vacuum

FIGURE 1.4.4b: The particle theory of matter explains how sound travels through solids, liquids and gases. Why does sound not pass through a vacuum?

- In a solid the particles are packed very closely together. Also, the forces between the particles are more elastic. The vibrating particles collide with neighbouring particles and bounce back very quickly, so the sound wave progresses very quickly.

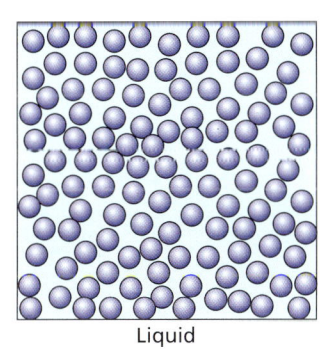

4. Draw a particle diagram with arrows showing how sound travels through a liquid.
5. Why do you think sound travels much faster through some solids compared to others?
6. Temperature can also affect the speed of sound. Develop a **hypothesis** to explain why.

Know this vocabulary

particle
vacuum
hypothesis

SEARCH: speed of sound in solids, liquids and gases

Waves

Learning about the reflection and absorption of sound

We are learning how to:
- Recognise which materials reflect the quality of sound.
- Analyse the effect of different materials on sound waves.
- Use ideas about energy transfer to explain how soundproofing works.

Concert halls are designed for good acoustics – so that the music sounds good to the whole audience. This means controlling the amount of echo and making sure sound reaches all corners. Different materials and shapes are used to achieve this.

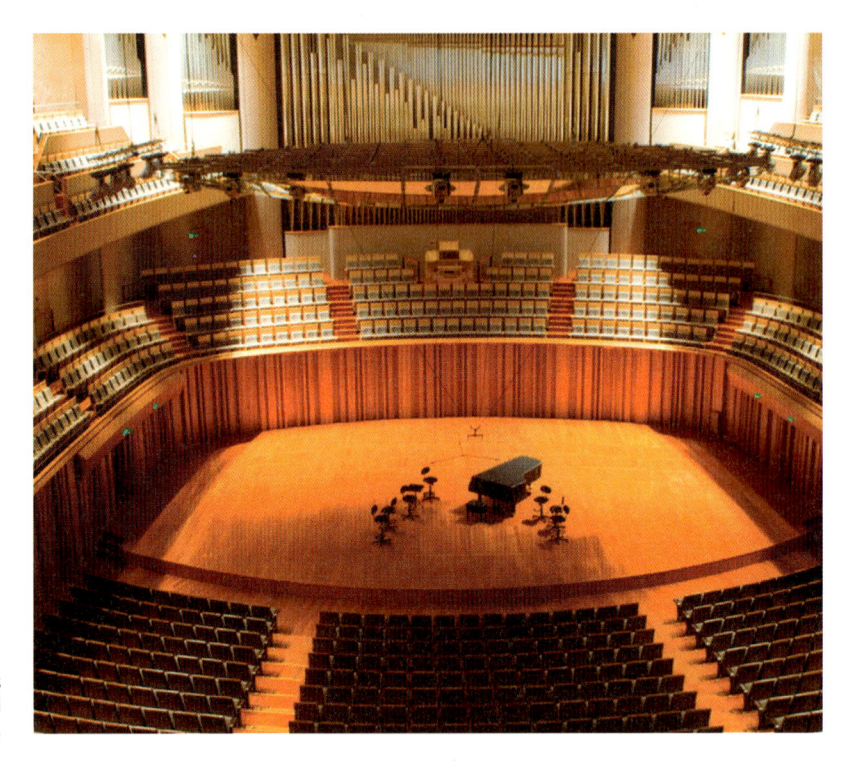

FIGURE 1.4.5a: Some materials help to reflect sound and others help to absorb it.

Effect of materials on sound waves

An **echo** is a sound wave that is reflected back to our ears. Hard, flat surfaces reflect sound well and produce strong echoes.

Soft surface materials that contain lots of air pockets, like fabric, foam and sponge, are not good at reflecting sound, but absorb it. This process is called **absorption**. The sound waves transfer energy to the air in the pockets so less is reflected.

1. What do we mean by 'absorbing' sound?
2. a) What would you hear if the sound waves from a bell were directed at a metal panel?
   b) What would you hear if the sound waves from a bell were directed at a panel made of sponge?
3. When might it be useful to absorb sound waves?

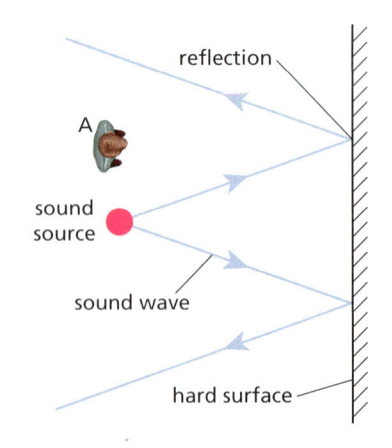

FIGURE 1.4.5b: Imagine you are standing in position A. What kind of echoes will this sound produce? How can the echoes be reduced?

86 AQA KS3 Science Student Book Part 1: Waves – Sound *and* Light

Effect of shapes on sound waves

4.5

Some materials can be shaped to reflect sounds in different ways. Look at the jagged surface in Figure 1.4.5c. When sound waves hit this surface, the reflected waves do not bounce back to the source. They are, instead reflected randomly, mostly away from the source.

The curved surface, on the other hand, reflects the sound until all the energy focuses towards a particular point. The sound at this point will be the loudest, whereas in places away from it, hardly any sound will be heard at all.

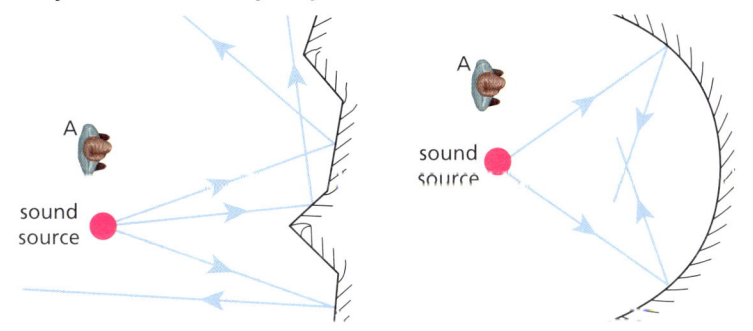

FIGURE 1.4.5c: How sound is reflected off a jagged and a curved surface.

FIGURE 1.4.5d: This material is used in soundproofing. What makes it a good choice?

4. Imagine you are standing in position A in each of the diagrams in Figure 1.4.5c. Describe what you will hear if the surface is:

   a) jagged   b) curved,

   compared with standing in position A, in Figure 1.4.5b, opposite a flat surface.

Soundproofing

When sound waves hit soft surfaces, they are absorbed by the air pockets. The sound waves become trapped, bouncing around in the air pockets, until all the energy is transferred as heat. Any sound reflected from the surface is therefore much quieter, as the sound waves have much less energy.

These soft materials are useful as **soundproofing**. A vacuum is also useful in soundproofing. Sheets of glass with a near-vacuum between them (very few gas particles) are very effective in stopping sound.

In the outdoor environment, trees, embankments and dense bushes are often used for soundproofing around mining areas.

5. What happens to the energy of the sound wave during absorption?
6. Suggest a soundproofing plan for a hospital in a busy town centre.

Did you know...?

A 'whispering gallery' is the name given to a large circular room, where a whisper made in one place is reflected to the opposite side of the room and heard there but nowhere else. St Paul's Cathedral contains one.

Know this vocabulary

echo
absorption
soundproofing

SEARCH: reflection and absorption of sound

Waves

Exploring properties of light

We are learning how to:
- Describe how light passes through different materials.
- Explain the difference between scattering and specular reflection.
- Explain how shadows are formed in eclipses.

Simple sundials can be made easily. The pointer is made of an opaque material that blocks light and produces a shadow. The position of the shadow can be used to tell the time.

FIGURE 1.4.6a: A sundial works because it casts a dark shadow.

See-through?

Light passes through gases, some liquids and some solids. Materials that light can pass through freely are said to be **transparent**. They do not produce shadows.

Other materials cast shadows by either completely or partially blocking the passage of light. **Opaque** materials block the passage of light waves completely, producing a dark shadow, whereas **translucent** materials only allow some of the light to pass through, casting weak shadows.

FIGURE 1.4.6b: Frosted (translucent) glass provides some privacy because it does not allow all the light to pass through it.

Transparent: some light is always reflected at the surface; most of the light is transmitted.

Translucent: some of the light is reflected; some of the light is absorbed; some of the light is transmitted.

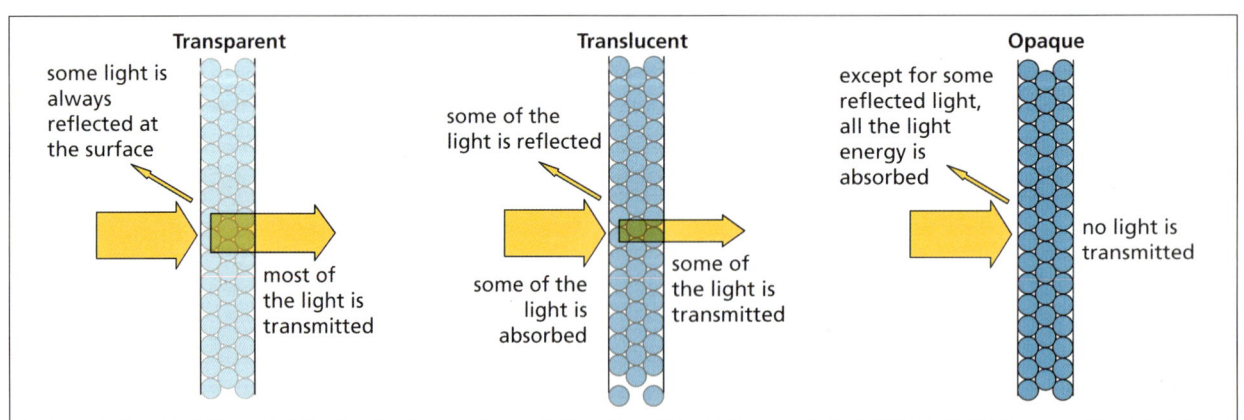

Opaque: except for some reflected light, all the light energy is absorbed; no light is transmitted.

FIGURE 1.4.6c: What happens to light when it falls on transparent, translucent and opaque materials.

1. Give three examples of transparent materials.
2. Compare the shadows produced by an opaque material and those by a translucent material.
3. Explain why an opaque object casts a shadow.

Did you know…?

You can see through a piece of frosted glass (make it 'see-through') simply by putting a piece of clear sticky tape on it.

88 AQA KS3 Science Student Book Part 1: Waves – Sound *and* Light

Solar eclipses

A solar eclipse happens when the Moon is between the Earth and the Sun. For a viewer on the Earth the effect is spectacular; for a few minutes the sky is plunged into darkness and a vivid ring of light is seen. However, it isn't the same for everyone.

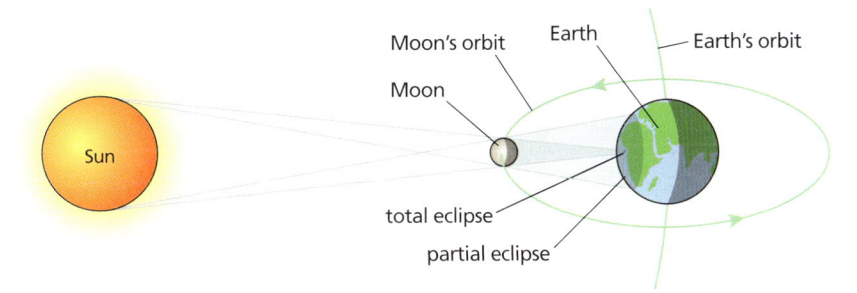

FIGURE 1.4.6d: The arrangement of the Sun, Moon and Earth during a solar eclipse.

The Moon's shadow isn't big enough to cover the whole of the Earth. Only people in a small area see a total eclipse. Just outside that, viewers experience a partial eclipse and in other places, nothing at all. This is because light travels in straight lines. The Moon is much smaller than the Sun so its shadow is smaller still.

4. What would you see if you were in an area of partial eclipse?
5. Why does the effect only last for a few minutes?

FIGURE 1.4.6e: A total eclipse is an amazing sight but you have to be in the right place.

Lunar eclipses

We also get an eclipse when the Earth blocks light from the Sun landing on the Moon. The three bodies are in alignment but this time the Earth is in the middle. The effect on Earth is quite different though; there is no shadow racing across the surface of the Earth but the Moon is bathed in a deep red light.

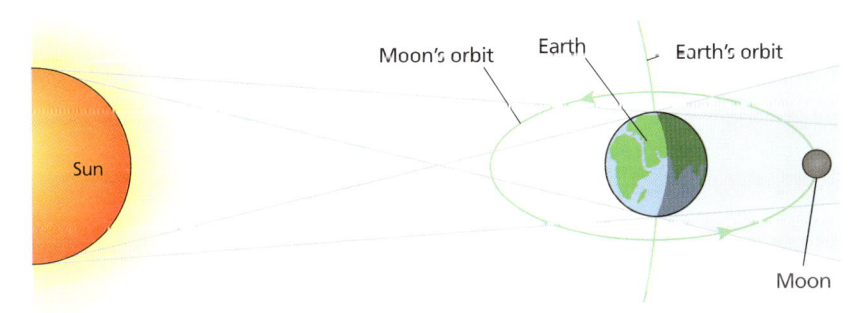

FIGURE 1.4.6g: The Moon being eclipsed.

FIGURE 1.4.6f: The arrangement of the Sun, Earth and Moon during a lunar eclipse.

6. Explain why lunar eclipses are only visible from certain parts of the Earth.
7. Draw and label a diagram to suggest whether lunar eclipses can be partial or total.

Know this vocabulary

transparent
opaque
translucent

SEARCH: light travel through transparent and opaque materials

Waves

Exploring reflection

We are learning how to:
- Describe how a mirror reflects light.
- Explain the difference between specular and diffuse reflection.
- Apply the law of reflection.

Light bounces off all kinds of surfaces – it's how we see the world around us. Often this reflected light scatters but if we use a mirror it follows a very particular route.

Explaining reflection

Light travels as waves. However, because a light wave travels in a straight line until it reaches a boundary, a **ray model** allows us to show clearly in a diagram the direction of the light and how it can change its direction when it meets a surface.

FIGURE 1.4.7a: Refraction causes the spoon to look as if it has broken into two parts.

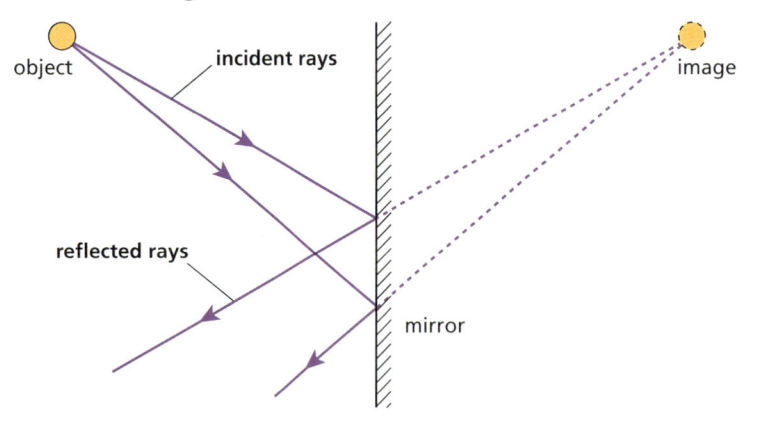

FIGURE 1.4.7b: The solid lines represent light rays. The dashed lines show where the light rays appear to be coming from – the image. This is the same distance from the mirror as from the object to the mirror.

1. Describe what Figure 1.4.7b shows.
2. Describe how the ray diagram showing image formation in a mirror needs to be changed if the object is:
   a) further away;   b) closer.

Specular and diffuse reflection

When light hits a surface or a boundary, some or all of it is reflected – it bounces back in a direction away from the surface. The reflection produced by a flat, smooth, shiny surface is called specular reflection. All the light is reflected in the same direction. It allows us to see an **image**. A rough reflective surface bounces light back in many directions. We can think of the surface as being a mixture of small flat surfaces at different angles. The effect is called diffuse reflection, or **scattering**.

FIGURE 1.4.7c: The specular reflection of light from a mirror allows you to see an image.

Did you know…?

An image ia a pictorial representation, usually two dimensional, of a physical object.

90   AQA KS3 Science Student Book Part 1: Waves – Sound *and* Light

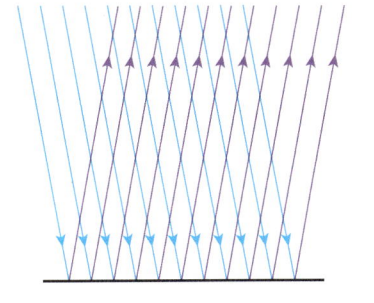

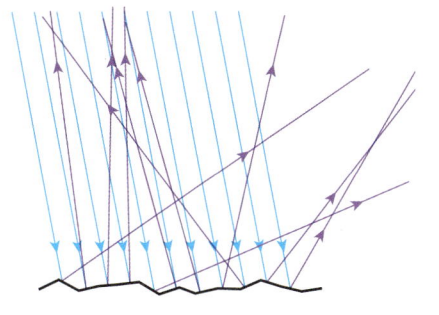

FIGURE 1.4.7d: The diagram on the left shows specular reflection. The one on the right shows diffuse reflection (scattering).

FIGURE 1.4.7e: The result of specular reflection in a calm lake.

3. Describe the difference between specular reflection and diffuse reflection.
4. Explain why scattering happens at rough surfaces.
5. Explain why reflections in lakes or ponds cannot be seen if the water is choppy.

Angles of reflection

If a ray of light is reflected by a mirror, the angle of the reflected ray is always exactly the same as the angle the ray came in at. As Figure 1.4.7f shows, the angles are not measured from the surface but from a line called the **normal line**, which is drawn at right angles to the surface. The incoming ray is called the incident ray and the outgoing ray is called the reflected ray. We can say:

angle of incidence = angle of reflection

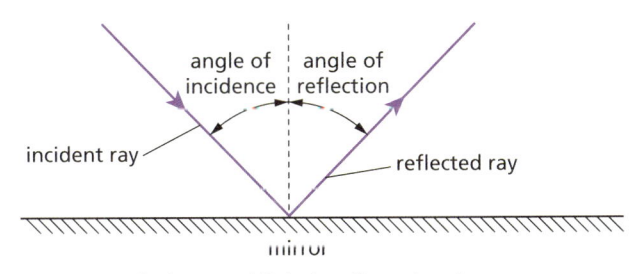

FIGURE 1.4.7f: The ray of light is reflected at the same angle it approached the mirror at.

FIGURE 1.4.7g

Did you know...?

Glass is capable of both reflecting light and letting it pass through – it depends on the angle. If you stand in front of a shop window you can see both what's in the shop but also reflections of things behind you.

6. What angle would light be reflected at if the angle of incidence were 20°?
7. What angle would light be reflected at if the angle between the incident ray and the surface of the mirror were 60°?
8. If we used a curved mirror, would the law of reflection still apply?
9. Copy Figure 1.4.7g and add two mirrors and rays to show how the person at A could see the person at B.

Know this vocabulary

ray model
incident ray
reflected ray
image
scattering
normal line
angle of incidence
angle of reflection

SEARCH: reflection of light

Physics

Exploring refraction

We are learning how to:
- Describe how light is refracted when it enters a different medium.
- Explain how this can cause it to change direction.
- Apply ideas about refraction to understanding lenses.

We usually think of light travelling in straight lines and usually it does. However, we can make it bend by making it pass through a transparent material that is either denser or less dense.

FIGURE 1.4.8a: Optical lenses.

Refraction

Light travels through transparent materials, such as glass and air. However, its speed depends on the density of the material. The denser the transparent material, the slower light travels through it. Glass is denser than air, so light travels through it slower than it travels through air.

Although we often draw light as rays, we need to remember that it is a series of transverse waves that transfer energy. Sometimes to explain what is happening we need to show the wave crests.

When the light slows down, the wavelength of its waves becomes shorter. Each of the crests catches up with the one in front. If the light then emerges into a less-dense medium it will speed up again. This effect is called **refraction**, and can cause the direction of the light to change.

Notice in Fig 1.4.8b the way the direction of the ray alters. When the ray goes into a denser medium (such as entering glass from air) it bends towards the normal. When it enters a less dense medium (such as entering air from glass) it bends away from the normal.

1. Describe what happens to light waves when they travel into a dense material.
2. Sketch a diagram to suggest what would happen if a light ray hit a glass rod as shown in Figure 1.4.8c.

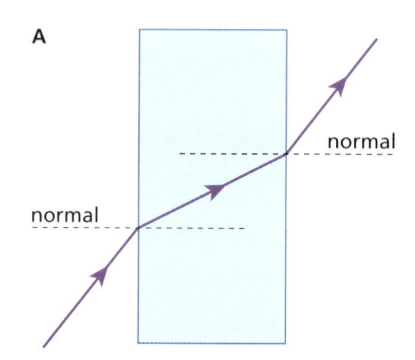

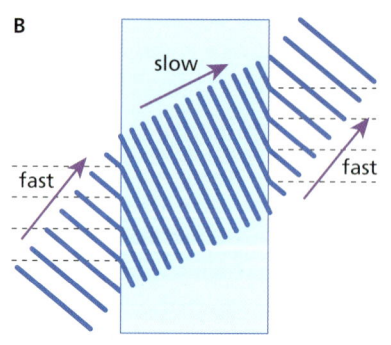

FIGURE 1.4.8b: A: ray model to show refraction of light; B: the wavelength shortens in a denser material.

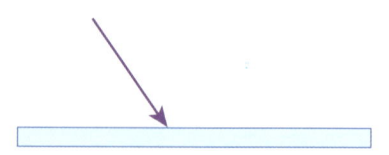

FIGURE 1.4.8c: A light ray hitting a glass rod – what happens?

Explaining how lenses work

Prisms come in different shapes but the most common ones are triangular. A ray of light passing through this will be refracted as it enters and as it leaves.

We can use ideas about refraction and prisms to explain how a **lens** works. Remember what happens when light is refracted – if it enters a denser medium it is refracted towards the normal and if it enters a less dense medium it will be refracted away from the normal. We can apply this to a triangular prism. Think about what happens when light enters it and when it leaves it.

Now consider the shape of a **convex lens**. We can think of it as being two prisms and a block (Figure 1.4.8d). It's not a perfect model but it's close enough. Look at the three parallel rays approaching. The upper one will be refracted downwards and the lower one upwards. The middle one will come through the middle and the three of them will meet. What the lens will then do is to bring rays together to produce a focused image. This can be caught on a screen.

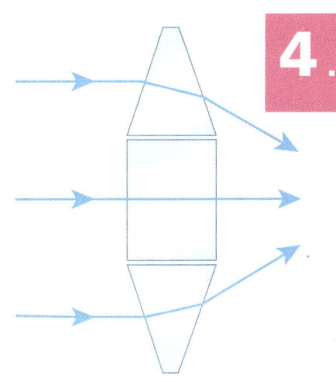

FIGURE 1.4.8d: We can simplify a convex lens to two prisms and a rectangular block to understand how it works.

3. Unless there's an object at the point where the rays meet in Figure 1.4.8d, they won't stop there. Sketch how they meet and where they will go next.
4. What difference do you think it will make if the convex lens is thinner? Show your ideas on a ray diagram.
5. Try constructing a ray diagram for a concave lens.

Did you know...?

The distance from a convex lens to the point where it makes parallel rays meet is called the focal length. The stronger the lens, the shorter the focal length.

Types of lens

Lenses are used in cameras, projectors, microscopes and telescopes. They come in many different shapes and sizes but there are two main types.

A convex lens bulges outwards in the middle. It makes light rays bend towards one another. A magnifying glass or hand lens is a common example of a convex lens. If we shine parallel light rays into a convex lens, they will meet each other at a focus and cross over. We sometimes call this a converging lens.

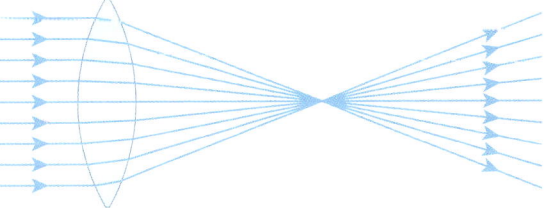

FIGURE 1.4.8f: A convex lens causes rays of light to converge.

A **concave lens** is thicker around the edge and curves inwards in the middle. It makes light rays spread out. If we shine parallel rays into a concave lens they will travel away from each other. This is sometimes called a diverging lens.

6. Suggest what a hand lens does with the light that passes through it from the object you are looking at.
7. Explain why:
   a) a convex lens is called a converging lens;
   b) a concave lens is called a diverging lens.

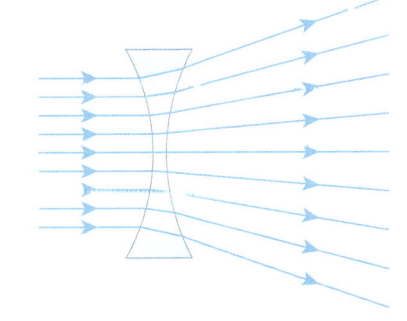

FIGURE 1.4.8g: A concave lens causes rays of light to diverge.

Know this vocabulary

refraction
lens
convex lens
concave lens

SEARCH: refraction of light

Waves

Seeing clearly

We are learning how to:
- Describe how the human eye works.
- Explain how the eye focuses on objects different distances away.
- Apply ideas about lenses to the correction of vision.

Our eyes have to cope with seeing things under many different conditions. Sometimes the object is close, or it may be far away, but we still need a clearly focused image. Our eyes are usually good at coping with this but may need a bit of help.

The human eye

Light enters an eye through the cornea and then travels through the lens. These both refract light rays, focusing them on a common point. An image forms on the **retina**. The optic nerve sends information to the brain, which interprets it.

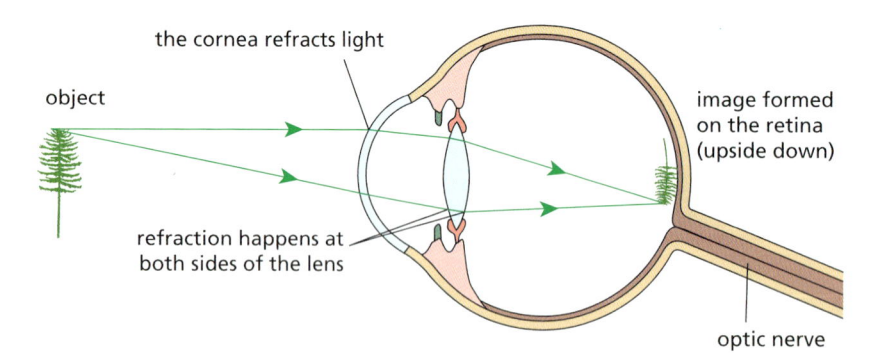

FIGURE 1.4.9b: How the eye works.

1. Name the parts of an eye that refract light.
2. Explain what the lens does to the rays of light.
3. Explain why the image formed on the retina is upside down.
4. Suggest why the lens in the human eye is a convex lens.

Looking at different objects

To see something, the rays need to be focused on the back of the eye. We need to be able to focus on objects that are close to us and also ones that are a long distance away. To do this, the lens in the eye changes shape.

For a distant object, a thin lens is sufficient to focus the rays on the back of the eye.

If an object is near, the light rays coming from it need to be refracted through a greater angle so the lens becomes fatter and more powerful.

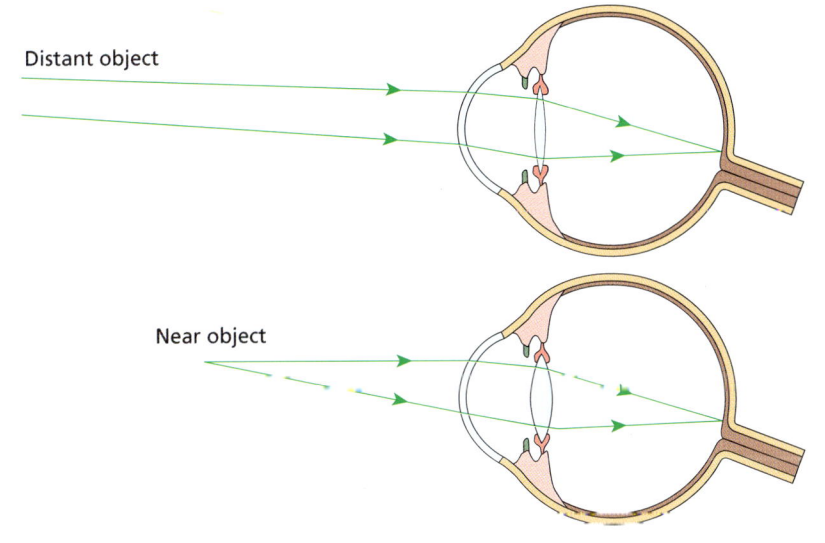

FIGURE 1.4.9c: The lens in the eye changes shape to focus on objects different distances away.

5. How is the lens in the eye different to the type of lens that would be used in spectacles?

6. Why do you think there's a limit as to how close an object you can focus on?

Correcting vision

Some people have eyes that won't focus properly on objects a certain distance away. What the lens in the eye should do is to focus rays on the back of the eye (on the retina). We'll look at two common defects. Look at Figure 1.4.9d.

One of these conditions is nearsightedness; the lens in the eye makes the rays meet before they reach the retina. The remedy is spectacles with concave lenses, which compensate by making the light rays diverge.

The other condition is farsightedness. This is the opposite situation; the lens in the eye doesn't make the rays close in sharply enough at the retina. The remedy is convex spectacle lenses.

7. If an object is near, light rays spread out from it and some may enter the eye. Draw a diagram to show how the eye will focus these and explain why it needs a strong lens.

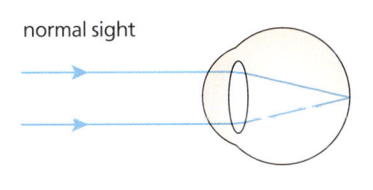

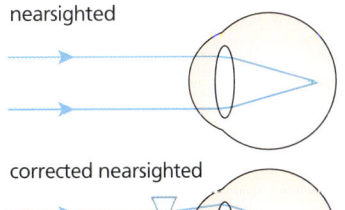

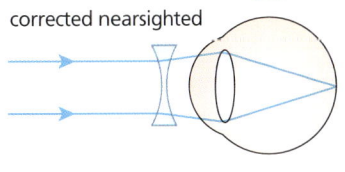

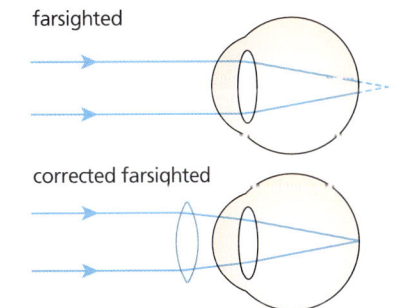

FIGURE 1.4.9d: How the lenses of glasses help to correct problems with vision.

Know this vocabulary

retina

SEARCH: how the eye works, correcting vision problems with lenses

Waves

Exploring coloured light

We are learning how to:
- Describe how a spectrum can be produced from white light.
- Compare the properties of light at different frequencies.
- Explain how light of different wavelengths can be split and recombined.

Some days it can be raining and, at the same time, the sun can be shining. This is when a rainbow can often be seen. To make a rainbow, sunlight and droplets of water are needed.

Spectrum from white light

Sunlight is made up of light waves of different **frequencies** and so different **wavelengths**. The range of wavelengths that the human eye can detect as different colours is called the visible **spectrum**. Seen together they make what is called white light. This white light can be split up to produce the colours of the spectrum. For example, if sunlight is shone through a triangular prism, it is refracted into different colours (Figure 1.4.10a and c).

1. Describe the spectrum obtained when white light passes through a triangular prism.
2. Which colours appear at each edge of a rainbow?

FIGURE 1.4.10a: A prism produces a spectrum from white light.

Different frequencies

In a vacuum all light travels at the same speed – 3000 million metres per second. Each colour of light has its own frequency. If light enters a denser medium – such as going from air into glass – it slows down, but the higher frequencies slow down more.

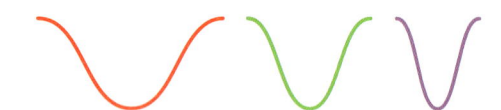

FIGURE 1.4.10b: In the visible spectrum, waves of red light have the longest wavelengths and waves of violet light, the shortest.

When a light wave passes into and out of a glass prism, the wave is refracted. The shorter its wavelength, the more it is refracted – so violet light is refracted more than red light. The 'white light' is split up and spreads out to form a spectrum.

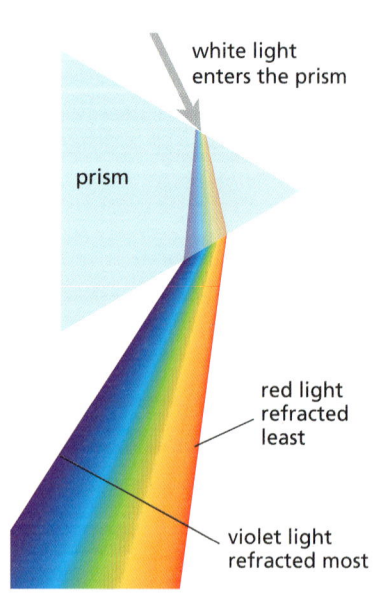

FIGURE 1.4.10c: The white light is split up by refraction.

96   AQA KS3 Science Student Book Part 1: Waves – Sound *and* Light

4.10

Waves with different wavelengths can be combined – this is additive colour mixing. It is different from mixing paints. Mixing red, blue and green light produces white light. Mixing red, blue and green paint produces muddy brown paint.

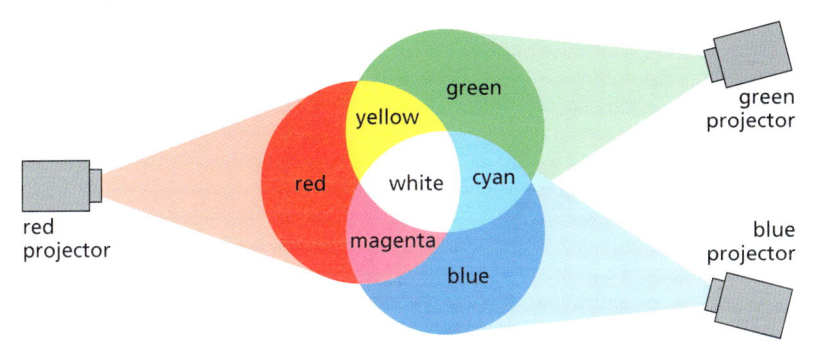

FIGURE 1.4.10d: Recombining waves to make white light.

3. Explain the relationship between the wavelength and the frequency of a wave.
4. Explain why white light spreads out to produce a visible spectrum when it passes through a prism.
5. If a white panel were lit with a red spotlight and a green spotlight, what colour would it appear to be?

Frequency and behaviour

Some materials are coloured, but you can see through them, such as coloured plastic sheets and solutions of food dyes. They absorb light waves of certain frequencies. The light that passes through consists of light with frequencies that were not absorbed.

Opaque coloured materials also absorb light waves of certain frequencies, but the light with other frequencies is reflected.

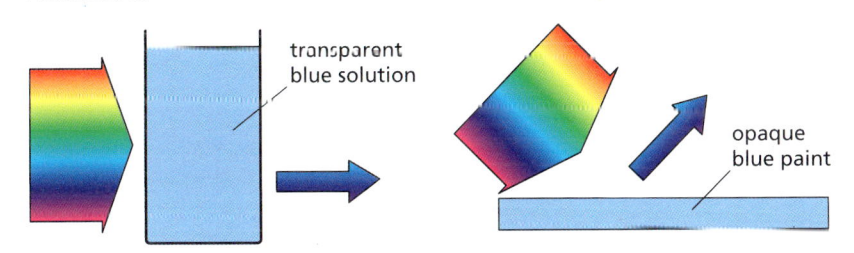

Did you know...?

Light with frequencies lower and higher than those at the extremes of the visible spectrum cannot be seen by the human eye. Infrared radiation has frequencies lower than that of red light. Ultraviolet radiation has frequencies higher than that of violet light.

FIGURE 1.4.10e: A blue solution absorbs wavelengths of light other than blue. Only blue light passes through. Blue paint absorbs wavelengths of light other than blue. Blue light is reflected. No light passes through.

6. Explain why a solution of red food colouring is red, but also transparent.
7. What colour would a green box seem to be if illuminated with:
   a) white light?
   b) red light?
   c) green light?
   d) blue light?

Know this vocabulary

frequency
wavelength
spectrum

SEARCH: light spectrum

Waves

Checking your progress

To make good progress in understanding science you need to focus on these ideas and skills.

- ☐ Recognise that sound energy is transferred by waves and describe how sound waves are made in different situations.
- ☐ Explain how longitudinal waves carry sound. Relate the terms frequency, wavelength and amplitude to sounds.
- ☐ Interpret and devise wave diagrams to represent different sounds of different frequency and amplitude.

- ☐ Know that sound consists of vibrations in a medium.
- ☐ Know that sound travels faster in some media than others.
- ☐ Understand that the denser the medium, the faster the sound travels.

- ☐ Recognise an echo as a reflection of sound.
- ☐ Recognise that some materials are good at reflecting sound and others can absorb it.
- ☐ Explain what is meant by reflection and absorption of sound.

- ☐ Know that sound can be represented by a waveform.
- ☐ Explain how the waveform represents the amplitude and wavelength of the sound.
- ☐ Interpret waveforms for different sounds.

- ☐ Understand that we hear sound because of vibrations travelling through a medium.
- ☐ Explain that we can hear a certain range of frequencies.
- ☐ Suggest how various ear problems might affect a person's hearing.

- ☐ Recognise that light can be reflected by some materials and absorbed by others.
- ☐ Explain the differences between transparent, translucent and opaque materials.
- ☐ Use diagrams to explain the difference between specular reflection and scattering.

4.11

☐ Describe the ray model of light, using the idea that light travels in straight lines.	☐ Explain the difference between reflection and refraction, and describe what happens when light waves are refracted.	☐ Use ray diagrams to explain reflection and refraction.
☐ Use the conventions of a ray diagram correctly.	☐ Use a ray diagram to show what happens when light is reflected.	☐ Use a ray diagram to show what happens when light is refracted.
☐ Recognise convex and concave lenses.	☐ Explain how convex and concave lenses affect light.	☐ Explain how lenses can be used to correct defects of vision.
☐ Describe the formation of a spectrum from white light.	☐ Explain how white light can be split into a continuous spectrum of colours, called the visible spectrum.	☐ Use the concepts of reflection and absorption of light to explain why some materials (transparent, translucent and opaque) are coloured.
☐ Explain how shadows are formed.	☐ Explain how solar and lunar eclipses occur.	☐ Explain why eclipses may be total or partial.
☐ Describe how light of different colours varies in terms of frequency.	☐ Explain how various colours can be obtained by using the three primary colours.	☐ Explain how the colour of an object is affected by the colour of light it is illuminated with.

Waves

Questions

KNOW. Questions 1–4

See how well you have understood the ideas in this chapter.

1. What range of frequencies of sound can most people hear? [1]
   a) Below 20 Hz
   b) Between 20 and 20 000 Hz
   c) Above 20 000 Hz
   d) There is no range

2. Draw a waveform which represents a loud, low-pitched (i.e. deep) note. Label it to show how you have represented those qualities. [2]

3. A light ray approaches a glass block at 30° to the normal. When it enters the block, which of these does the ray do? [1]
   a) Bend towards the normal
   b) Continue travelling in the same direction as before
   c) Bend away from the normal
   d) Travel along the normal

4. How does sound travel in a vacuum compared with in air? [1]
   a) It travels faster in a vacuum
   b) It travels at the same speed as in air
   c) It travels slower in a vacuum
   d) It won't travel at all in a vacuum

APPLY. Questions 5–9

See how well you can apply the ideas in this chapter to new situations.

5. Look at the different waves shown in 1.4.12a. Wave a) represents a note played in the middle of a piano. Which wave best represents a siren? [1]

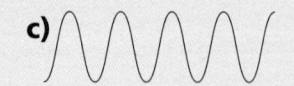

   a)   b)   c)   d)

   FIGURE 1.4.12a

6. Emily's family are moving house. Their lounge is empty, with no curtains, carpets or furniture, and it sounds 'echoey'. Which of these statements is correct? [1]
   a) Hard surfaces are good at absorbing sound.
   b) Sound travels faster than light.
   c) Sound travels faster in an empty room.
   d) Soft surfaces such as curtains are good at absorbing sound.

7. Think about what happens to sunlight when it passes through transparent materials, and then explain why we see different colours in stained glass windows. [2]

8. Draw and label diagrams to show the different effects that concave and convex lenses have on parallel rays of light. [1]

9. Explain whether you would expect sound to travel through water faster, slower or at the same speed as in air, and why. [2]

EXTEND. Questions 10–12

See how well you can understand and explain new ideas and evidence.

10. Ultrasound is high-frequency sound, beyond the auditory range of humans. A bat sends out an ultrasound signal. It receives an echo just 0.5 seconds later. How far away is its prey? (Distance = speed × time; the speed of sound in air is 330 m/s.) [2]

11. Imagine looking at a small object through a block of glass. Complete a copy of Figure 1.4.12b to show where the object appears to be (its image). [1]

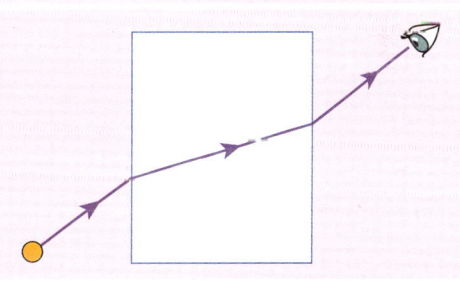

FIGURE 1.4.12b

12. A periscope has an arrangement of mirrors to enable the user to see around obstacles. Complete the diagram to show:

    a) how light rays travel from the object to the user;
    b) that the object does not appear upside down. [3]

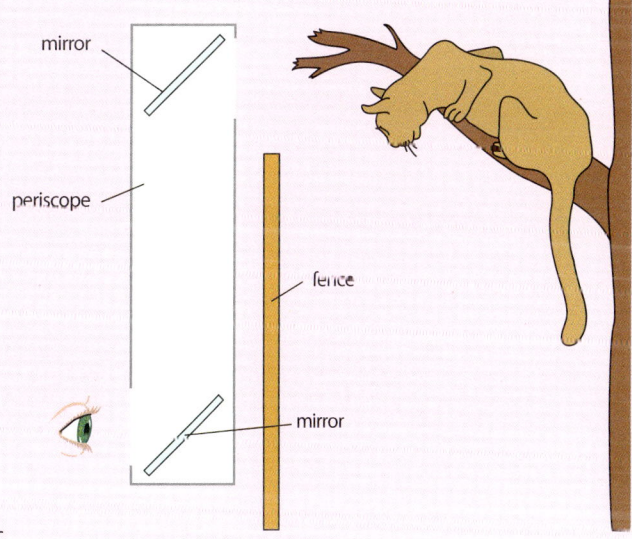

FIGURE 1.4.12c

Matter
Particle model *and* Separating mixtures

Ideas you have met before

States of matter

Solid, liquid and gas are the three main states of matter, and most materials can be grouped into one of these.

When materials are heated or cooled, they may change from one state to another. Water freezes to become ice at 0 °C, and boils to becomes a gas at 100 °C.

In the water cycle, water evaporates to become a gas, condenses in clouds and forms water droplets. It falls back to Earth as precipitation.

Reversible changes

Physical changes, such as changes of state, are reversible. Water can be frozen to make ice; this can melt to form liquid water.

Dissolving and mixing are also reversible changes – salt can be added to water, which can be evaporated to recover the solid salt.

Dissolving and solubility

Some materials – such as salt and sugar – can dissolve in water. We say that these are soluble and the mixture forms a solution.

Other materials – such as sand – do not dissolve in water. We say that these are insoluble.

5.0

In this chapter you will find out

Using the particle model

- The particle model explains why solids have a fixed shape and cannot flow, and why liquids and gases do not have a fixed shape, and can flow.
- Particles in solids, liquids and gases have their own internal energy – the energy of particles in a gas is far higher than the energy of particles in liquids and solids.
- The effect of temperature can be explained using the particle model. This explains how changes of state take place and how solids, liquids and gases expand on heating.
- We can also explain differences in density, concentration and pressure using the particle model. These differences can account for why perfume spreads in a room.

Separating mixtures

- If solid material has been mixed with water but has not dissolved, we can separate it by using a filter or a sieve.
- If we heat a liquid it will evaporate, turning into a vapour (gas). If we then cool the vapour, it will turn back into a liquid. This process is called distillation.
- We can use information about different boiling points to separate mixtures of liquids. Distillation is used to make perfume and also fuels such as petrol.
- Soluble substances can be made to travel up filter paper by adding a solvent.
- If we do this with coloured dyes or inks, we find that the different colours in the mixture move different distances.
- This technique is called paper chromatography and can be used to separate mixtures and identify chemicals.

Matter

Using particles to explain matter

We are learning how to:
- Recognise differences between solids, liquids and gases.
- Describe solids, liquids and gases in terms of the particle model.

Have you ever wondered why it is possible to put your hand through a liquid such as water, or a gas, such as air, but not through a solid wooden door? The answer lies in how the particles are arranged in these states of matter.

Particle arrangement

Anything that takes up space and mass is called 'matter'. All matter is made from **particles**. Particles vary in the ways they are arranged and behave. These are known as different states of matter. Figure 1.5.1a uses a **particle model** to show how particles are arranged in the three most common states of matter – solids, liquids and gases.

1. Name three solids, three liquids and three gases you are familiar with.
2. Describe how the arrangements of particles in solids, liquids and gases differ from each other.

Particles and internal energy

All particles above the temperature known as absolute zero (−273 °C) have internal **energy**. Particles in solids, liquids and gases have different amounts of energy.

- In solids, the particles vibrate in their fixed positions.
- Particles in liquids move randomly from their positions, but are always in contact with other particles.
- Particles in a gas move about randomly and very fast, widely separated from but colliding with other particles.

Temperature affects how fast particles move. At higher temperatures, particles in a solid vibrate faster, while in liquids and gases particles move around faster.

3. Draw a cartoon to describe how the energies of the particles in solids, liquids and gases vary.
4. In which of the following do the particles have the most internal energy – ice, oxygen at room temperature, or steam (over 100 °C)?

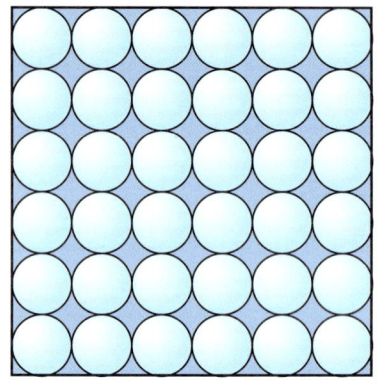

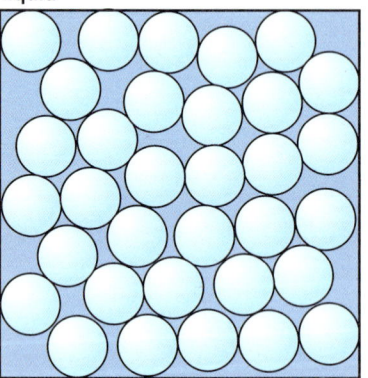

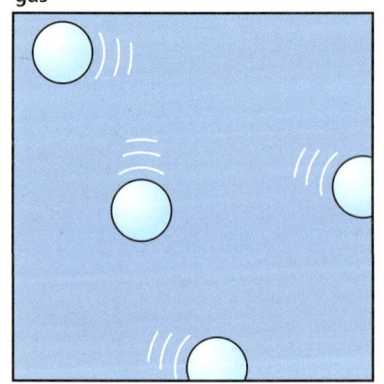

FIGURE 1.5.1a: Solid – particles are in fixed, regular positions.

Liquid – particles are close together and touching. They can move from their position.

Gas – particles have no fixed position and are far away from each other. They can move very fast.

104 AQA KS3 Science Student Book Part 1: Matter – Particle model *and* Separating mixtures

Intermolecular forces

5.1

The particles in a solid have very strong, attractive **intermolecular forces** between them, which hold the particles in their positions. Between particles in liquids, the intermolecular forces are still strong, but not as strong as in a solid. This is why the particles are able to move about. The intermolecular forces between the particles of a gas are very weak.

Density is a measure of how much matter there is in a particular volume. The stronger the intermolecular forces are, the more matter can fit into a volume and, therefore, the more dense the substance is.

FIGURE 1.5.1b: How do the intermolecular forces in this solid ice make it rigid?

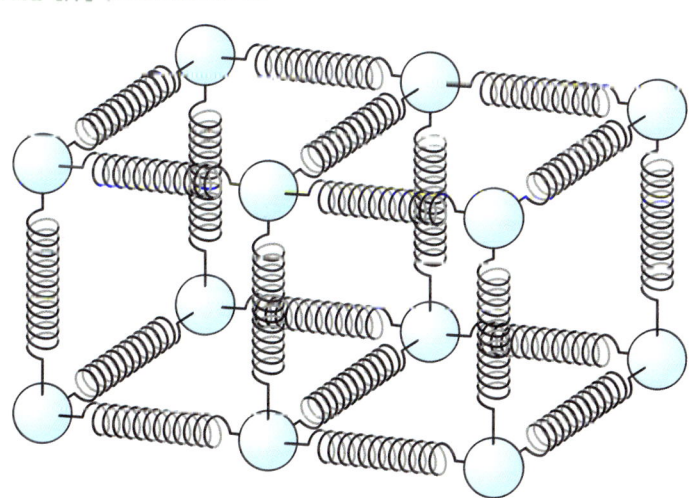

FIGURE 1.5.1c: Forces between particles can be represented by springs. How would you modify this particle model to show a solid with strong intermolecular forces and one with weak intermolecular forces?

Some solids, like metals, have very strong intermolecular forces between the particles – others, like paper, are not nearly as strong.

5. Use ideas about intermolecular forces to explain why you can put your hand through air but not through wood.

6. What can you say about the intermolecular forces between the particles of jelly compared with those of a metal?

7. Compare the density of jelly and the density of solid metal. Explain your answer.

8. Describe the relationship between the energy of the particles and the intermolecular forces holding them together.

Did you know…?

The most common state of matter in the Universe is called 'plasma'. It is known as the fourth state of matter, and is a form of gas. The Sun and space are made of plasma. We can make tools from plasma to cut strong metals.

Know this vocabulary

particle
particle model
energy
intermolecular forces
density

SEARCH: particle model   105

Matter

Understanding solids

We are learning how to:
- Describe the properties of solids.
- Relate the properties and behaviour of solids to the particle model.

Some properties are common to all solids and help to define them. Differences between solids can be explained using the particle model.

General properties of solids

Except for mercury, all metals are solid at room temperature. They have high melting points and boiling points, and all conduct heat and electricity well. A few non-metallic solids share these properties, but many others have very different properties.

Flow

Some solids appear as if they can flow, like sand. Seen under a microscope, such a solid is made up of many individual grains. None of the matter in the individual solid grains can flow.

Changing shape

Some solids are **malleable** – they can be hammered into shape without being broken. Solids such as plastic are **brittle** – they will snap if hit. Metals are **ductile** – the layers of particles are able to slide past each other, so they can be pulled into extremely thin wires.

Strength

Strength is the ability of a solid to withstand a force. Metals are generally very strong.

Hardness

Hardness is a measure of how easy it is to scratch a solid – it is not the same as strength. Slate and concrete are very strong solids, but are easily scratched so they are not hard.

Solubility

Some solids, salt for example, dissolve readily in water – they are **soluble**. Others, such as sand, do not dissolve.

Conduction of heat and electricity

Metals will readily **conduct** heat and electricity, whereas **non-metal** solids, like plastic and rubber, will not. The only exception is graphite (a form of carbon), which conducts electricity even better than metals.

FIGURE 1.5.2a: The hardest and strongest material in the world.

Did you know...?

Diamond is the strongest, hardest material in the world. Drills that cut through rock are tipped with diamond.

TABLE 1.5.2: Put these solids in order of hardness.

Substance	Hardness
aluminium	3
carbon (diamond)	10
iron	4
silver	2.5
tin	1.5
copper	3

5.2

1. Which property makes copper a good choice for making wires?
2. Use the data in Table 1.5.2 to explain which material you would use on the end of a drill.

What are alloys?

Alloys, such as brass, bronze and chrome, are mixtures of metals. They are often stronger than the individual metals they are made from. Different sizes and colours can be used to represent the different types of atoms in the particle model, as in Figure 1.5.2b.

3. Duralumin is an alloy made from 96 per cent aluminium and 4 per cent copper. What might the particle arrangement look like?
4. Use the particle model to explain why some alloys are less ductile than the metals they are made from.

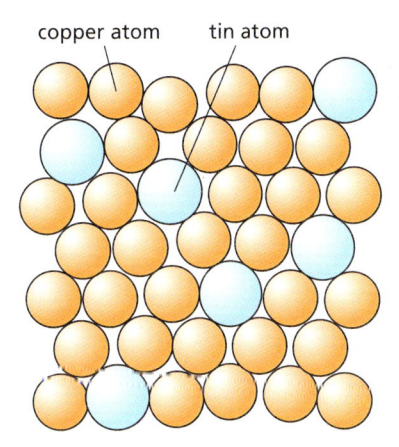

FIGURE 1.5.2b: Bronze is made up of 85 per cent copper and 15 per cent tin

Explaining properties

Strength, shape, density and hardness all depend on the strength of the intermolecular forces between the particles in a solid.

Solubility also depends on intermolecular forces. In solids which dissolve, forces between the particles of the solid are weaker than the forces between the particles of the solid and the particles of the liquid.

The arrangement of particles in metals is special, accounting for their ability to conduct heat and electricity. This is shown in Figure 1.5.2c.

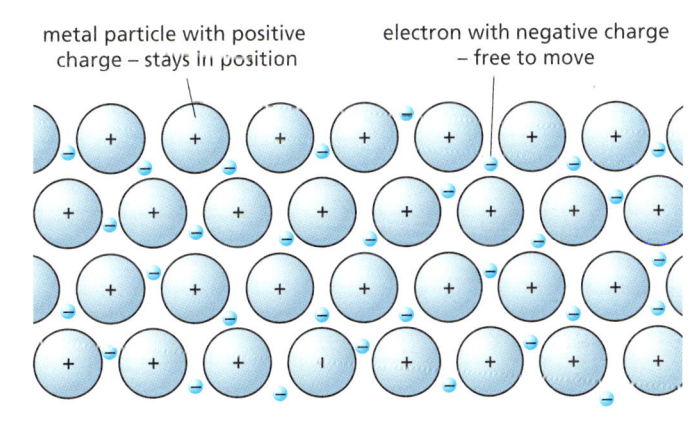

FIGURE 1.5.2c: Metals can conduct heat and electricity because they have small particles (negatively charged electrons) that move freely.

5. How would you modify the particle model to show a solid with strong intermolecular forces and one with weak intermolecular forces?
6. Explain as many differences in the properties of copper and wax as you can, using appropriate particle models.

Know this vocabulary

- malleable
- brittle
- ductile
- strength
- hardness
- soluble
- conduct
- non-metal
- alloy

SEARCH: properties of solids

Matter

Understanding liquids and gases

We are learning how to:
- Describe the properties of liquids and gases.
- Relate the properties and behaviour of liquids and gases to the particle model.

We rely on the properties of liquids and gases every day. For example, we rely on the compression of gases to fill a car or bike tyre and in our cans of hairspray or deodorant. Properties of liquids and gases can be explained using the particle model.

Viscosity

Liquids and gases can be poured and can flow. This is because the intermolecular forces holding the particles of a liquid in place are quite strong, but not strong enough to keep the particles in position. They are able to slide over and roll around each other.

Some liquids flow more easily than others. Liquids in which the intermolecular forces are stronger do not flow as easily because it is more difficult for the particles to slide past each other. Resistance to flow is known as **viscosity**. Think of oil and water – oil flows more slowly than water; it is more viscous.

1. Give three applications of liquids and gases that rely on their ability to flow.
2. You want to oil your bike. You decide to investigate three different brands to find out which spreads most easily in the cold.
    a) Which variables must you control?
    b) What would you measure?

The effects of compression

Compression is the process of squashing a material so that the particles move closer together. In a solid and a liquid, there is very little space and, therefore, it is not possible to compress them. In a gas, the particles have space between them and so they can be pushed together more closely. Gases can be compressed.

FIGURE 1.5.3a: Which of these materials is the most viscous?

Figure 1.5.3b shows cylinders that have been filled with gas compressed so much that it has become a liquid. A can of hairspray or deodorant is a smaller version of this. The particles of gas have been forced so close together that they are touching and now are arranged as in a liquid.

3. Explain why oxygen gas is more easily compressed than water or ice.
4. Consider the different substances and compare and explain the particle model for each:
   a) liquid nitrogen and nitrogen gas;
   b) ice, water and steam;
   c) deodorant when it is in the can and as it is sprayed into the air.

Explaining pressure in gases

When gas particles move, they have collisions between themselves and also with the sides of the container they occupy. The **gas pressure** is a measure of the average force of these collisions over the area of the container's sides. The standard units of pressure are **kilopascals** (kPa).

Pressure is increased when there are more particles and, therefore, more collisions with the sides of the container.

FIGURE 1.5.3b: The cylinders contain gas which has been liquefied. When released, the change in pressure causes the liquid particles to become gas particles again.

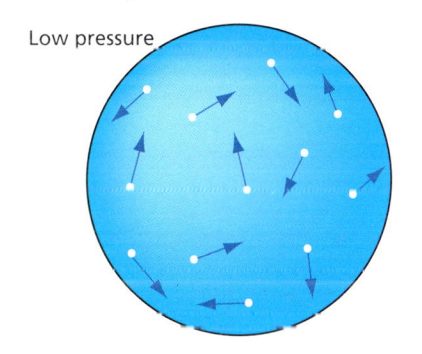

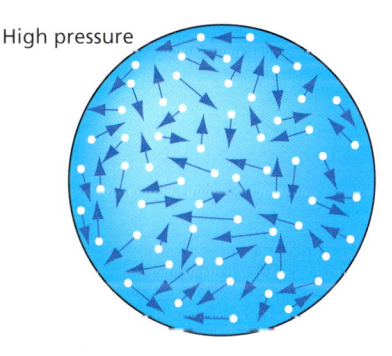

FIGURE 1.5.3c: What do we mean by pressure?

Increasing the temperature causes the particles to move faster. This increases the force of the collisions against the sides of the container, and their frequency, meaning that the pressure is increased.

5. Suggest one reason why atmospheric pressure is lower at the top of a mountain than at sea level.
6. Using the particle model, draw the same gas at three pressures: low, medium and high pressure.

Did you know…?

In the world's longest-running experiment, scientists have been measuring the viscosity of pitch tar by timing how long it takes to fall through a funnel. It has taken between 7 and 13 years for each drop to fall and the experiment has been running since 1927!

Know this vocabulary

viscosity
compression
gas pressure
kilopascal

SEARCH: properties of liquids and gases

Matter

Exploring diffusion

We are learning how to:
- Use the particle model to explain observations involving diffusion.

Diffusion is a process in which particles move and spread out. Unsurprisingly, gas particles diffuse much faster than particles in other states of matter. What makes diffusion so special?

Examples of diffusion

Diffusion occurs because of the movement of particles in a gas or a liquid. There is hardly any diffusion in solids because the particles cannot move freely. Gas particles move faster and further than liquid particles, so diffusion in gases occurs faster than in liquids.

All smells spread as a result of diffusion. When particles of a gas, like air freshener spray or odours from smelly socks, are released into the air, they spread out. These gas particles move through the air – when they reach your nose they are detected as a smell. This is why we can detect smells from a long distance away.

1. Give another example of diffusion in everyday life.
2. Why do smells become weaker the further you are from the source?

FIGURE 1.5.4a: If a drop of coloured ink is added to water, after several hours the colour will have spread through the water so that it is of equal concentration throughout.

Diffusion and the particle model

Concentration is a measure of the number of particles packed in a certain volume.

Diffusion occurs because particles move from an area of high concentration to an area of low concentration, until the concentration is equal throughout. We call this the point of **equilibrium**. The difference in concentration is known as the **concentration gradient**. The higher the concentration gradient, the greater the rate of diffusion.

Temperature affects the rate of diffusion because it affects the energy of the particles. The higher the temperature, the higher the kinetic energy of the particles, and the faster they move in such a way as to reduce the concentration gradient.

Did you know...?

The animal kingdom is full of amazing examples of how animals make use of diffusion to smell odours. Elephants can detect water sources from up to 20 kilometres away.

5.4

3. If a drop of ink is added to some pure water, and a similar drop of ink is added to some dilute ink solution, in which solution would diffusion happen fastest? Explain your answer.

4. Think about these examples of diffusion. Suggest which will reach equilibrium first and explain your answer.
   a) placing a spoonful of cordial in 50 cm^3 of hot water;
   b) adding a spoonful of coffee to 50 cm^3 of cold water.

Explaining diffusion

Look at Figure 1.5.4b. Concentrated hydrochloric acid is placed at one end of the tube and concentrated ammonia at the other end. Particles of ammonia are smaller than particles of hydrochloric acid. When the particles diffuse, they meet and react, forming a white cloud of ammonium chloride.

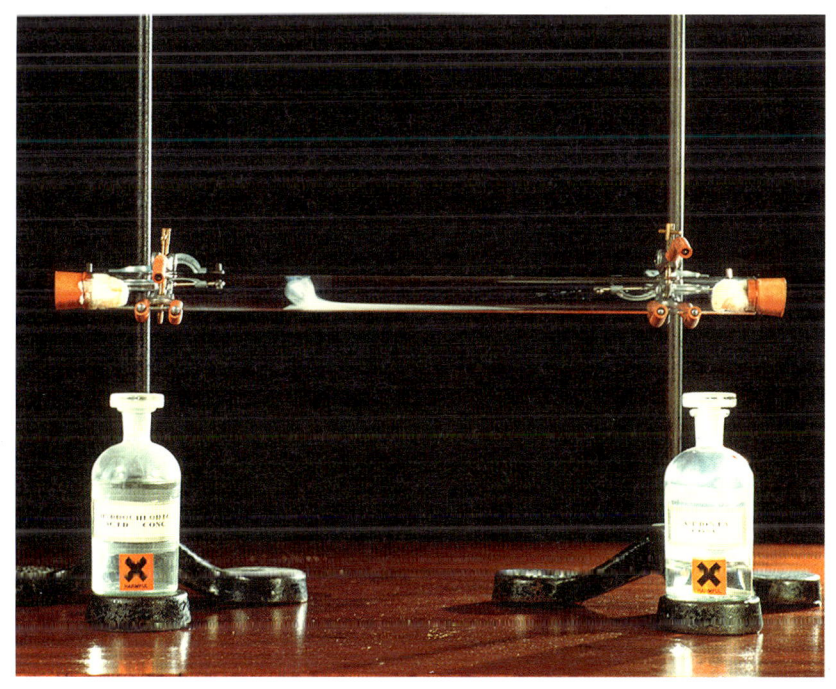

low

medium

high

FIGURE 1.5.4b: How do the hydrochloric acid particles and ammonia particles reach each other to react?

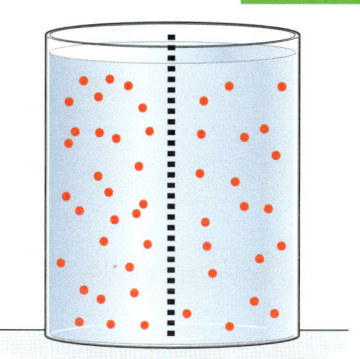

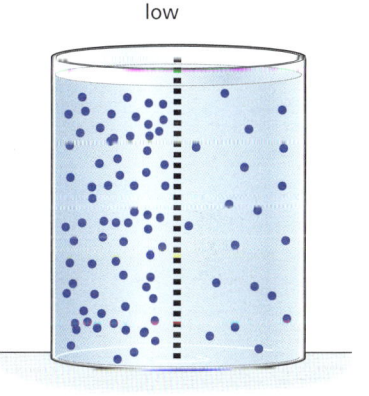

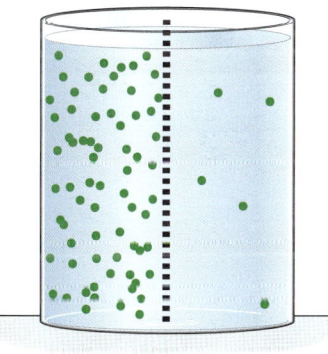

FIGURE 1.5.4c: Compare the concentration gradients in the diagrams. This affects the rate of diffusion.

5. What would happen if the concentrated solutions were replaced by dilute solutions of both hydrochloric acid and ammonia?

6. How might the formation of the white ring be speeded up? Explain your answer.

7. Why doesn't the white ring in Figure 1.5.4b form in the centre of the tube?

Know this vocabulary

diffusion
concentration
equilibrium
concentration gradient

SEARCH: diffusion 111

Matter

Explaining changes of state

We are learning how to:
- Recognise changes of state as being reversible changes.
- Use scientific terminology to describe changes of state.
- Explain changes of state using the particle model and ideas about energy transfer.

When you make ice or **melt** the frost from a windscreen, you are making use of changes of state. What is actually happening to the particles in these processes?

Reversible changes

Have you ever seen 'dry ice'? It is solid carbon dioxide that is turning straight into a gas – there is no liquid state. This is a process called sublimation. Iodine is another example of a substance that **sublimes**. If the gas is cooled sufficiently, it turns directly into a solid.

Turning solids into liquids or gases, and liquids into gases are reversible changes. They are called physical changes.

Figure 1.5.5b summarises the processes by which substances change their state.

FIGURE 1.5.5a: Solid carbon dioxide is known as 'dry ice'.

energy transferred to the particles from the surroundings by heat

solid → melting → liquid → boiling → gas
sublimation (solid → gas)
gas → condensation → liquid → freezing → solid

energy transferred from the particles to the surroundings by heat

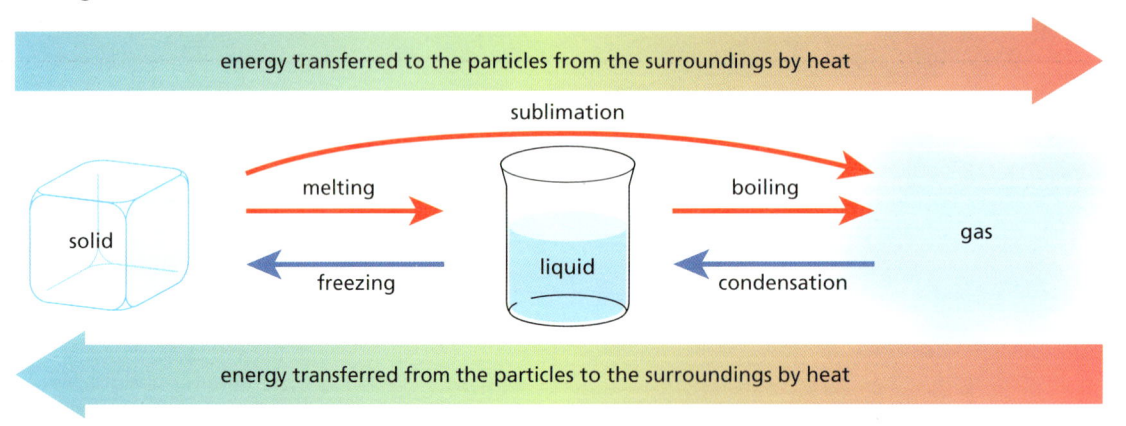

FIGUE 1.5.5b: The changes of state.

1. Describe how you could show that making water **freeze** is a reversible change.
2. Use Figure 1.5.5b to describe the meaning of the following words:
   a) melting;
   b) condensing;
   c) boiling;
   d) freezing;
   e) sublimation.

112   AQA KS3 Science Student Book Part 1: Matter – Particle model *and* Separating mixtures

Changing state

5.5

The temperature at which a pure substance melts or freezes is fixed – it is called the **melting point** or freezing point, depending on the change taking place.

When a pure substance **boils** or condenses, this also occurs at a fixed temperature called its **boiling point**.

Different substances have different melting points and boiling points. These points depend on the strength of their intermolecular forces.

> 3. Aluminium melts at 660 °C but copper melts at 1064 °C. Explain why, in terms of intermolecular forces.
> 4. At 0 °C, hydrogen is a gas, mercury is a liquid and water is a solid. What can you infer about the inter-particle forces in each from this data? Explain your answer.

Did you know…?

Helium has the lowest melting point of all elements at −272 °C, whereas diamond (carbon) has the highest melting point, at 3500 °C.

Differences between boiling and evaporation

At the upper surface of a liquid, the liquid **evaporates**. Evaporation occurs at any temperature between the melting point and the boiling point. It occurs only at the surface of the liquid. Some of the surface particles gain enough kinetic energy from heat, from the surroundings, to leave the surrounding particles in the liquid and become a vapour. Over time, all the particles at the surface will evaporate.

FIGURE 1.5.5c: How is evaporation different from boiling?

Some liquids, such as alcohol, have weaker intermolecular forces than others, such as water. Therefore, with the same amount of energy transferred from the surroundings, more alcohol particles than water particles will escape as a gas.

Boiling, however, occurs only at the boiling point, and the whole liquid changes into a gas. The water particles gain sufficient energy from their surroundings to leave the liquid.

> 5. Is evaporation more likely to take place at temperatures near the boiling point or near the melting point? Explain your answer.
> 6. What are the main differences between boiling and evaporation?

Know this vocabulary

melt
sublime
freeze
boil
condense
melting point
boiling point
evaporate

SEARCH: changing states of matter

Matter

Separating mixtures

We are learning how to:
- Recognise the differences between substances and use these to separate them.

If you put different objects together, such as different fruits in a bowl, toys in a box or sweets in a bag, you have a mixture. Air, fruit juice, milk and sea water are also mixtures. Some mixtures are harder to separate than others.

Using size to separate mixtures

A **pure substance** contains only one type of particle – for example, gold is a pure substance. A **mixture** is made up of at least two pure substances – for example, gold and copper together make a mixture.

Gravel and rocks can be removed from sand by sieving. This separation depends on the size of the holes in the sieve. However, if the sand were mixed with water, this method would not work. A **filter** would be needed instead.

Filters are often made of paper or cloth with very small holes that are difficult to see without a microscope. Filters are often used to remove the solids when making coffee. Tea bags act as filters, whereas a tea strainer acts as a sieve. Air and fuel filters are used in cars to remove particles that would damage the engine.

1. What is the difference between a filter and a sieve?
2. Explain how filters and sieves are helpful when making tea and coffee.

FIGURE 1.5.6a: Using a sieve.

Being different

Mixtures can be separated by finding differences in physical properties between the substances. For example, there are only three metals that are attracted by a magnet – iron, cobalt and nickel. We can use this difference to separate these **magnetic** metals from mixtures.

FIGURE 1.5.6b: The physical property of magnetism can be used to separate magnetic from non-magnetic materials.

114   AQA KS3 Science Student Book Part 1: Matter – Particle model *and* Separating mixtures

5.6

Rules for mixtures
1. Mixtures can be separated by physical methods.
2. Mixtures only have the properties of the things in the mixture.
3. Mixtures of substances can be made using different amounts of each one.
4. No chemical change occurs when making mixtures.

TABLE 1.5.6

3. Choose a method to separate flour and rice.
4. Would all of a mixture containing iron filings and lead powder be magnetic?
5. If nickel chloride were mixed with lead, could you use a magnet to separate them? Explain your answer.
6. Use the rules in Table 1.5.6 to explain why mixtures can be separated using known differences between the substances.

Did you know…?

The components of blood can be separated by spinning the blood really fast in a centrifuge. This causes the red blood cells to separate from the plasma because they are denser and move to the bottom.

Separation by filtration

In the laboratory, filter paper can be used to separate some solids from liquids – this process is called **filtration** (Figure 1.5.6c). Substances that do not dissolve in a liquid are described as **insoluble**. Filtration separates out these insoluble solid substances.

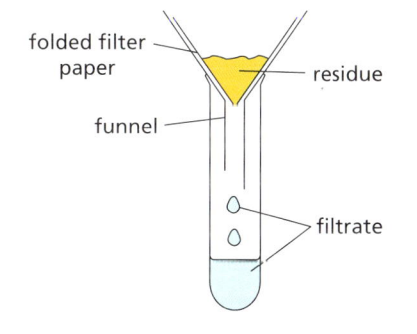

FIGURE 1.5.6c: Separating mixtures by filtration.

Liquids like oil and water do not mix. The oil does not dissolve in the water to make a solution. These liquids are described as **immiscible**. The lighter oil floats on top of the water, and even if you shake the mixture, the two layers will reappear as the two liquids separate again.

The way these two liquids behave means that a separating funnel can be used to split them up (Figure 1.5.6d). The water layer can be removed using the tap at the bottom, leaving the oil layer behind.

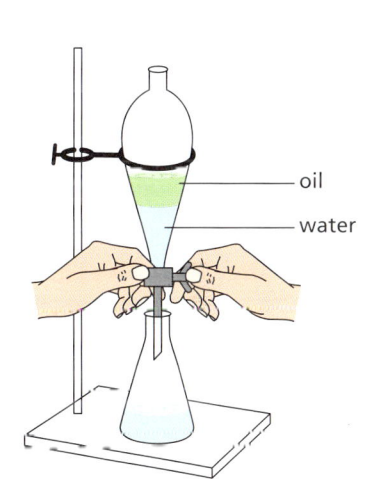

FIGURE 1.5.6d: Using a separating funnel to separate immiscible liquids.

7. Explain why a bottle of salad dressing made from vinegar and olive oil must be shaken before use.
8. Explain why filtration would not separate sugar and water.
9. Create a key or flow diagram to help explain which method of separation to use for a mixture of your choice.

Know this vocabulary

pure substance
mixture
filter
magnetic
filtration
insoluble
immiscible

SEARCH: separating mixtures

Matter

Exploring solutions

We are learning how to:
- Explain the terms solvent, solution, solute and soluble.
- Separate a soluble substance from water.
- Analyse patterns and present data to explain solubility.

Limestone caves are amazing places. Stalactites grow down from the roof and, where the water drips down and hits the cave floor, stalagmites grow upwards. They are formed as minerals that were once dissolved in the water come out of solution and form solid deposits.

FIGURE 1.5.7a: Even rocks dissolve.

Do you take sugar?

If you stir sugar into a cup of tea or coffee the crystals disappear – they **dissolve**. The water is called the **solvent**, the sugar is the **solute** and the mixture is called a **solution**. The sweeter the taste, the more sugar has dissolved. Substances that dissolve are described as **soluble**.

One way to help things dissolve is to increase the temperature of the water. This is why we wash clothes in warm water. Any soluble stains in the clothes will dissolve better at a higher temperature. The mass of solute that dissolves in a solvent at a particular temperature is called its **solubility**.

Look at the data in Table 1.5.7. The results show the mass of sugar (sucrose) that can dissolve in 100 g of water.

TABLE 1.5.7: Dissolving sugar in water at different temperatures.

Temperature of water (°C)	0	20	40	80
Mass of sucrose that can dissolve (g)	180	200	240	600

Did you know…?

These amazing natural gypsum crystals were found 300 metres underground in a mine in Mexico. They have grown undisturbed for thousands of years. Some are as long as 12 metres.

FIGURE 1.5.7b: Naica gypsum crystals.

1. What is a solution?
2. What does the data in Table 1.5.7 tell you about the solubility of sucrose at different temperatures?
3. How could you display this data to show the pattern more clearly?
4. Estimate the mass of sucrose that will dissolve in 100 g of water at 60 °C.

Using graphs

5.7

Soluble substances dissolve more easily in hot water because the water molecules have more energy and move faster. They can break down the solute crystals and separate the solute particles more quickly.

Solubility also depends on the type of solute. The graph in Figure 1.5.7c shows the change in solubility of different salts with temperature.

> 5. Look at Figure 1.5.7c. Which salt is most soluble at 60 °C?
> 6. If 50 g of potassium nitrate were added to water at 20 °C, would it all dissolve? How do you know?
> 7. Using your knowledge of dissolving, explain why there is a connection between the temperature of a solvent and the solubility of a salt.

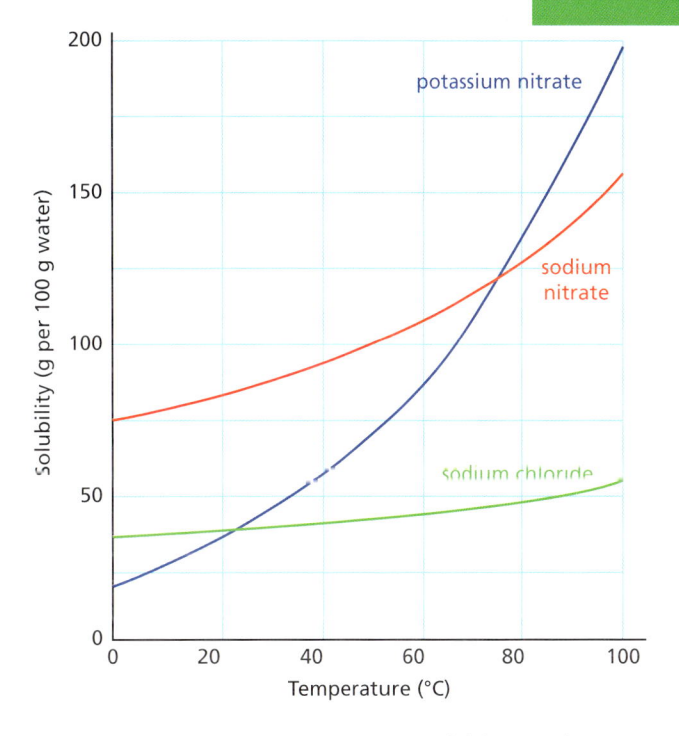

FIGURE 1.5.7c: Solubility graphs.

Explaining solubility

Some substances are more soluble than others. Solute particles have forces of attraction between themselves and the solvent particles. When the forces of attraction between the solute and solvent are stronger than the inter-particle forces between the solute particles, the solute will dissolve. The solute particles fill the spaces between the solvent particles. When all the spaces are filled up, the solution is said to be saturated, because no more solute can be dissolved.

If the temperature is increased, the solvent particles move with more energy, moving further apart and creating more space. More solute can be dissolved at the higher temperature because there are more gaps that can be filled.

> 8. Adapt the particle model to explain why some solids are more soluble than others.
> 9. Can the particle model be used to show how temperature affects solubility? What are the strengths and limitations of the model?

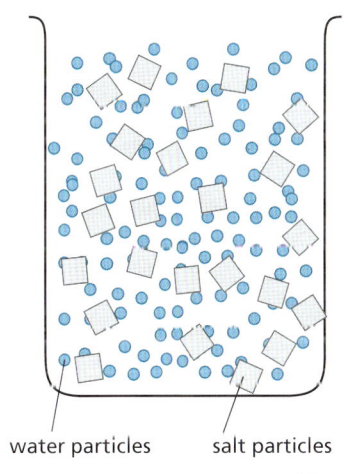

FIGURE 1.5.7d: Solute particles fill the spaces between the solvent particles while dissolving.

Know this vocabulary

dissolve
solvent
solute
solution
soluble
solubility

SEARCH: solutions and solubility

Matter

Understanding distillation

We are learning how to:
- Use distillation to separate substances.
- Explain how distillation can purify substances.

Distillation is used in making perfumes, fuels (such as petrol) and alcoholic drinks (such as vodka). It is an important separation process involving heating and cooling.

Heating and cooling

On a cold day water **vapour** from a bath or kettle can **condense** on a cold surface. It cools down and turns back to water. This is one of the processes involved in **distillation**. Liquid mixtures can be separated using distillation.

When water boils it is hard to catch all of the water vapour because it mixes into the air. In distillation the vapour is cooled, which allows it to be collected as a liquid.

The distillation apparatus that we use is shown in Figure 1.5.8b. In the Liebig condenser, the hot vapour from the boiling liquid flows through the inner tube, while cold water runs through the outer tube. This keeps the inner glass tube cold and condenses most vapours easily. The liquid collected at the end is called the distillate.

FIGURE 1.5.8a: Condensation is one of the processes involved in distillation.

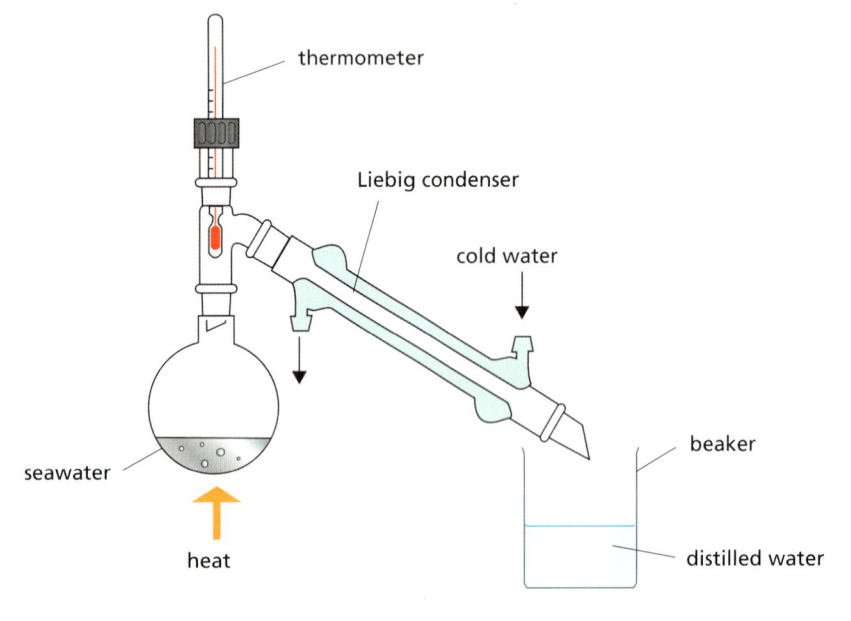

FIGURE 1.5.8b: Distillation apparatus using a Liebig condenser.

Did you know...?

Steam distillation is used to obtain essential oils from plants such as herbs and flowers. The products are used in aromatherapy, flavourings in foods and drinks, and as scents in perfumes, cosmetics and cleaning products.

5.8

1. Why does steam turn into liquid water when it touches a window?
2. Describe the structure of a Liebig condenser.
3. Explain why the Liebig condenser uses cold water.

Distilling mixtures

There are two changes of state in distillation. First, a liquid is evaporated by heating and then the cooled vapour is condensed back to a liquid. When salty water is heated, only the water (solvent) changes state and the salt (solute) is left behind. The water produced is called distilled water.

Different liquids boil at different temperatures – for example, ethanol boils at 78 °C and water at 100 °C. This means that mixtures of liquids can be separated using distillation. A thermometer at the top of a distillation flask shows the temperature of the vapour being condensed and hence identifies the substance being separated. Distillation is an effective way of **purifying** ethanol or increasing the concentration of alcoholic drinks. It is also useful for separating flammable liquids like petrol and diesel because the vapours never come into direct contact with the flame.

4. Name three substances that are separated using distillation.
5. Why is a thermometer important in distillation?
6. Explain how water and ethanol are separated.

The challenge of separating

When separating mixtures, the properties of the substances in the mixture must be considered.

For example, salt can be separated from a solution of salt water using evaporation because water evaporates but the salt doesn't. However, if we wanted to preserve the water as well, we would need to use distillation.

A mixture of solids may be separated by using a magnet if only one of the mixture is magnetic, for example, iron, cobalt or nickel.

7. Consider the mixtures that could be made from the pairs of substances in Figure 1.5.8c. Describe how you would separate them, and explain your reasoning.

ink and water

sugar cubes and icing sugar

saltwater and sand

iron filings and copper turnings

FIGURE 1.5.8c: How would you separate mixtures of these pairs of substances?

Know this vocabulary

vapour
condense
distillation
purify

SEARCH: distillation

Matter

Exploring chromatography

We are learning how to:

- Use chromatography to separate dyes.
- Use evidence from chromatography to identify unknown substances in a mixture.

Chromatography is one of the most important separation methods used to identify unknown substances. There are many types of chromatography – some use liquids and some use gases. Chromatography is used by scientists to detect drugs and explosives and to identify dyes and paints.

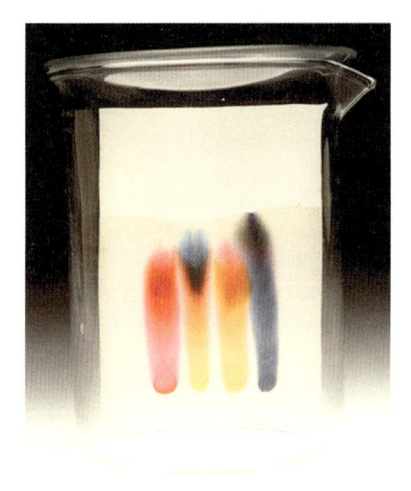

FIGURE 1.5.9a: Separation by chromatography.

Separating colours

Black ink is not just a black colour mixed with water. Black ink is a mixture of colours. Filter paper and water can be used to separate these colours. This method of separation is called paper **chromatography**. Figure 1.5.9b shows a simple method. Drops of water (solvent) are added to the middle of the paper where the ink spot is placed.

1. What evidence is there that black ink is not pure?
2. What causes the ink to spread across the wet filter paper?
3. Describe what we mean by chromatography.

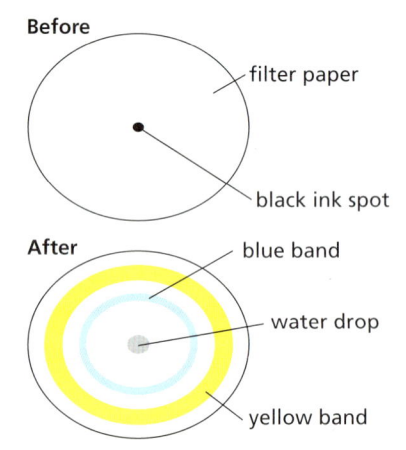

FIGURE 1.5.9b : A simple method of separating ink colours.

Examples of chromatography

If you cut a section of the filter paper, it can act as a wick. By dipping this wick into water, the liquid is drawn up through the ink and the colours begin to separate.

This method, shown in Figure 1.5.9c, is called ascending paper chromatography, because the water soaks up from the base, carrying the colour spots with it. Some colours move faster than others, which is why the colours separate.

The resulting pattern of colours is a **chromatogram**.

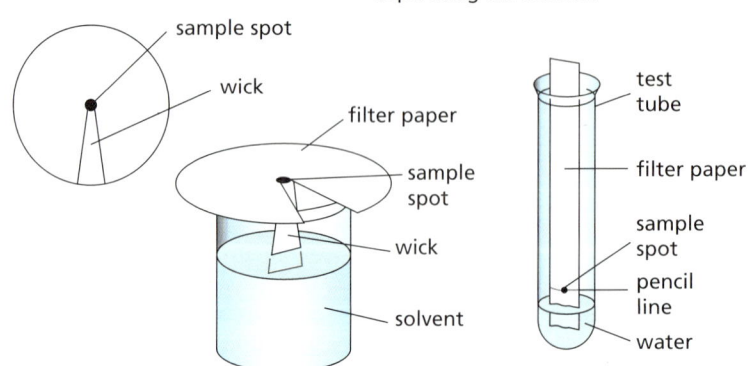

FIGURE 1.5.9c: Another method of paper chromatography.

120 AQA KS3 Science Student Book Part 1: Matter – Particle model *and* Separating mixtures

It is possible to use chromatography for colourless mixtures, but the chromatogram must be developed by spraying the paper with a chemical to make the spots visible, or using an ultraviolet (UV) light to look at the spots.

5.9

4. What do we call the pattern of colours on the paper?
5. In Figure 1.5.9c, why is the line drawn in pencil and not in ink?
6. Why would paper chromatography be no good for separating salt from water?

A special separation technique

Samples of DNA gathered at crime scenes can be used to identify or eliminate suspects. DNA is the material in our cells that we inherit from our parents. The sample is treated with special chemicals and then injected into a gel. When an electric current passes through the gel, the components of the DNA separate and spread, just like the ink on the chromatography paper. This is called electrophoresis. The pattern that the DNA produces is unique to an individual person, like a fingerprint.

Scientists can use DNA 'fingerprints' to find out who you are related to. Your DNA fingerprint contains aspects of the DNA patterns of each of your parents.

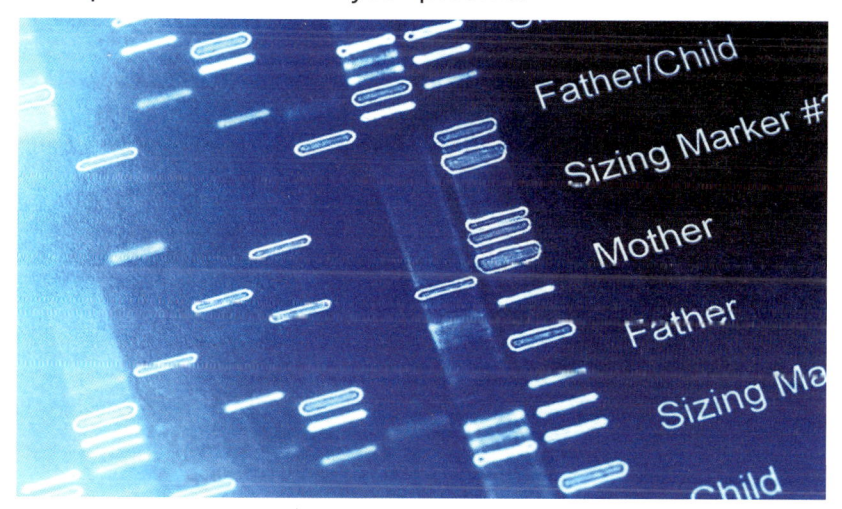

FIGURE 1.5.9e: DNA fingerprinting can help to incriminate suspects or rule them out.

7. Explain how DNA fingerprinting is similar to chromatography.
8. What are the differences between chromatography and electrophoresis?
9. What precautions would forensic scientists have to take when gathering and testing DNA evidence?

Did you know...?

You can separate pigments in leaves by chromatography using ethanol as the solvent. Chlorophylls are green pigments that help plants make food via photosynthesis. The yellow pigment separated here is carotene, which is found in carrots and is used as a food colouring (E160a).

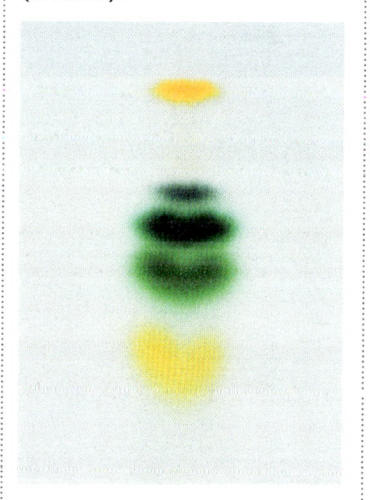

FIGURE 1.5.9d: A plant pigment chromatogram.

Know this vocabulary

chromatography
chromatogram

SEARCH: chromatography

Matter

Checking your progress

To make good progress in understanding science you need to focus on these ideas and skills.

☐ Compare the properties of solids, liquids and gases.	☐ Draw particle diagrams to demonstrate the differences between the arrangement of particles in solids, liquids and gases, and describe their different properties.	☐ Use particle diagrams to explain the differences in energy and forces between the particles in different states of matter, accounting for differences in their properties.
☐ Use correct terminology and the particle model to describe changes of state, including evaporation.	☐ Interpret data relating to melting and boiling points.	☐ Explain data relating to melting and boiling points in terms of intermolecular forces.
☐ Describe what is meant by the terms 'concentration' and 'pressure'.	☐ Describe the effects of changing concentration and pressure in terms of particles.	☐ Explain the effects of changing concentration and pressure in terms of particles, and apply to processes such as diffusion and gas compression.
☐ Describe some methods to separate mixtures.	☐ Select and explain appropriate separation techniques.	☐ Explain the choice and method of separation using correct terms.
☐ Define solvent, solute, solution and soluble.	☐ Interpret solubility graphs to compare solubility of different solutes and describe the effect of temperature on solubility.	☐ Explain solubility and the effect of temperature in terms of particles and intermolecular forces.

122　AQA KS3 Science Student Book Part 1: Matter – Particle model *and* Separating mixtures

5.10

- [ ] Describe the process of distillation.
- [ ] Explain the physical processes involved in distillation.
- [ ] Identify the uses and advantages of distillation.
- [ ] Identify mixtures using chromatography.
- [ ] Explain how to separate a mixture using chromatography.
- [ ] Use chromatograms to explain the composition of mixtures; compare chromatography and DNA analysis.

Matter

Questions

KNOW. Questions 1–7

See how well you have understood the ideas in this chapter.

1. Which of the following statements is true? [1]
   a) Particles in a solid have more energy than particles in a liquid.
   b) Particles in a gas have weaker intermolecular forces than particles in a liquid.
   c) Particles in a liquid have more internal energy than particles in solids and gases.
   d) Particles in a solid do not have any internal energy because they do not move.

2. What does diffusion depend on? A difference in: [1]
   a) temperature   b) state   c) concentration   d) mass

3. Describe, using ideas about particles, how temperature affects the viscosity of liquids. [2]

4. When steam hits a cold window, it becomes a liquid. Name this change of state of matter and explain what happens in terms of particles. [2]

5. What do we mean by an insoluble material? [1]
   a) It will not dissolve.
   b) You cannot get it back after it has dissolved.
   c) It dissolves other things well.
   d) It will dissolve easily.

6. Filtration separates mixtures on the basis of a difference in which property of the substances in the mixture? [1]
   a) Solubility   b) Boiling point   c) Magnetism   d) Particle size

7. Describe how you could safely separate salt from salt water, retaining both the salt and the water. [4]

APPLY. Questions 8–13

See how well you can apply the ideas in this chapter to new situations.

8. Iron has a melting point of 1535 °C and a boiling point of 2750 °C. At which temperature will iron be a liquid? [1]
   a) 2752 °C   b) 2751 °C   c) 1534 °C   d) 1536 °C

9. The density of water is 1 g/cm³ and that of syrup is 1.3 g/cm³. Which of the following statements is false? [1]
   a) In a mixture of syrup and water, the water will float on top.
   b) The syrup is more dense than the water.
   c) In a mixture of syrup and water, the syrup will float on top.
   d) There is more mass per unit volume in syrup than in water.

5.11

10. Figure 1.5.11a shows the particles of different substances. Which particle diagram represents a shaving foam aerosol? [1]

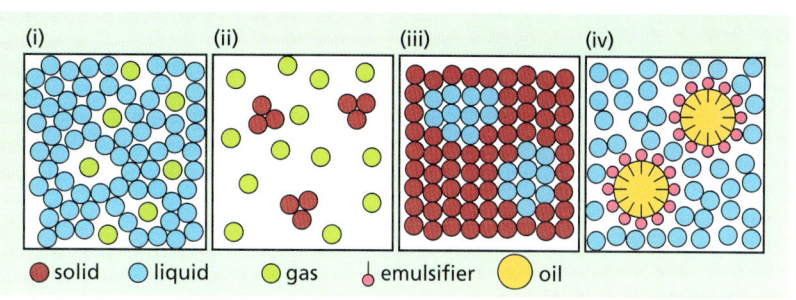

FIGURE 1.5.11a ● solid ○ liquid ○ gas ⁀ emulsifier ○ oil

11. Aerosol cans have a warning on them to prevent their use on fires. Explain reasons for this, using ideas about particles. [4]

12. A good way of separating a mixture of petrol and water is: [1]

    a) distillation;   **b)** filtering;   **c)** crystallisation;
    d) chromatography;   **e)** using a separating funnel.

13. Marcus and Lisa are investigating how well different brands of sugar dissolve in water. Using Table 1.5.11:

    a) Explain what the results show. [2]
    b) What do they have to do to make sure it is a fair test? [2]

Brand of sugar	Mass used (g)	Volume water (cm³)	How many stirs to dissolve it all
Kyle & Tait	10	100	14
Finegrade	10	100	8
Spoonful	20	200	12
Delightful	20	200	7

TABLE 1.5.11

EXTEND. Questions 14–15

See how well you can understand and explain new ideas and evidence.

14. Butane (C_4H_{10}) is camping fuel. Its boiling point is –1 °C. Hydrogen (H_2) is also a fuel, with a boiling point of –252 °C. Both fuels are transported under pressure, turning them into a liquid so that more particles can be carried. Explain which fuel will be easier to transport in this way and why. [2]

15. A vet has been asked to find out if any of four horses, A, B, C and D, have been drugged. She takes urine samples from the four horses and arranges for a lab to prepare a chromatogram to test the samples. What do the results show? [4]

Analysis of urine samples
1. Drug X
2. Drug Y
3. Horse sample A
4. Horse sample B
5. Horse sample C
6. Horse sample D

FIGURE 1.5.11b: Chromatogram of horse urine samples.

Reactions
Metals and non-metals *and* Acids and alkalis

Ideas you have met before

Metals

Materials can be grouped based on their properties such as hardness, solubility, conductivity and response to magnets.

Many useful materials, including plastics, wood and metals, have uses that exploit their properties.

Metals are shiny solids that we use for many applications, such as making cars, computers, bridges and so on.

Metals are good electrical conductors, which is why we use them to make wires for circuits.

Chemical changes

Changes can occur when materials are mixed. Some of these changes are non-reversible – these are called chemical changes, or chemical reactions.

Mixing bicarbonate of soda with vinegar or making toast are chemical changes – you cannot get the original materials back.

The new materials made in chemical changes can be useful.

Burning

Burning materials (such as wood, wax and gas) produces new materials.

Burning is a chemical change. Burning is also known as combustion.

In this chapter you will find out

6.0

Properties of metals and non-metals
- Most metals are solid and strong. Alloys often have different properties from their component metals, giving them different uses.
- Many non-metals are unreactive gases at room temperature.
- Some metals and non-metals have unusual properties, for example mercury and bromine.

Types of reactions
- We can represent reactions using equations and particle diagrams.
- Many metals react with acids to produce a salt plus hydrogen gas.
- Oxidation is a reaction with oxygen to form an oxide compound. Combustion and rusting are examples of oxidation.
- More reactive elements will remove less reactive elements from their compounds. This is known as displacement. We can use displacement reactions to predict a reactivity series.
- The reactivity series is a list of elements (mainly metals) arranged in order of their reactivity.

Acids, alkalis and indicators
- We use acids in our everyday lives, for example in food and batteries.
- We use alkalis in our everyday lives, for example in cleaning products and medicines.
- Some acids and alkalis are hazardous.
- We can make and use indicators to show how acidic or alkaline a substance is.
- The pH scale is an important measure of the level of acidity and alkalinity of a substance.

Reactions of acids and alkalis
- Acids react with metals and with alkalis.
- In these reactions the particles are rearranged – we can show this using diagrams, equations and other models.
- A neutral substance is one with pH 7. It is made when an acid and an alkali exactly neutralise one another.
- Neutralisation reactions can be useful for our health.

Reactions

Using metals and non-metals

We are learning how to:
- Recognise the properties and uses of metals and non-metals.
- Explain the uses of metals and non-metals based on their properties.

Some **metals** have similar properties, such as being strong and shiny. These properties help us in different uses. **Non-metals** are neither strong nor shiny, and some are gases at room temperature. However, some **metals and non-metals** are more unusual, for example, three metals are magnetic and one metal is a liquid at room temperature.

FIGURE 1.6.1a: Cars rely on different properties of metals.

Metals and their properties

We use metals for building because they are strong and for making jewellery because they are shiny and attractive. *Strong* and *shiny* are two properties of metals.

We use metals in electrical circuits because they all **conduct** electricity and are **ductile**, which means they can be stretched into wires. Metals are also **malleable**, meaning that they can be bent, rolled into sheets and shaped without them breaking. Most metals make a ringing noise when hit – we say that they are **sonorous**. They also conduct heat very well and most have high melting points.

1. List the properties common to most metals.
2. Which properties of metals are most important for making:
   a) saucepans?   b) water pipes?   c) drinks cans?

> **Did you know…?**
>
> Mobile phones contain more than 10 different metals including some of the rarest. The battery contains copper, cobalt, zinc and nickel. The circuit board and touchscreen may contain copper, gold, arsenic, cadmium, lead, nickel, silver, zinc, mercury, indium and tantalum.

Metals and their alloys

Iron is a very strong, grey metal which makes it useful as a structural material. Copper is an orange-coloured metal that is more malleable and ductile than iron. It is used in electrical circuits, wires and water pipes. Water pipes were made of lead until it was found to be harmful to living things. Unlike iron, copper does not corrode.

Iron, cobalt and nickel are unusual because they are the only **magnetic** metals. Magnetic materials have many uses, such as in toys, on cupboard doors and in credit cards. Other metals have unusual properties, for example, mercury is a liquid at room temperature and sodium is soft and is very reactive.

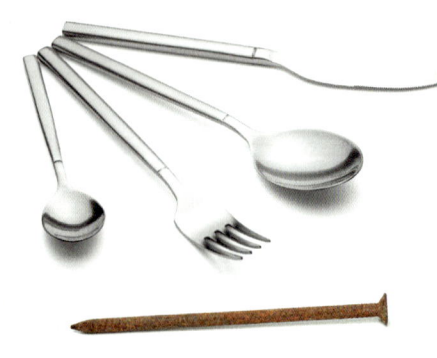

FIGURE 1.6.1b: Stainless steel is an alloy of iron that doesn't rust.

128   AQA KS3 Science Student Book Part 1: Reactions – Metals and non-metals *and* Acids and alkalis

6.1

Metals can be mixed together to form **alloys**. Alloys have different properties compared with the metals that they are made from, which sometimes makes them more useful. Stainless steel is an alloy of iron – adding different metals, like chromium, to the iron makes it stronger, shinier and less likely to rust.

> 3. List three metals in everyday use and state their uses. Explain how their properties make them useful.
> 4. Suggest why copper is not used in credit card strips and why sodium is not used in building.
> 5. What is an alloy and why are alloys often used instead of pure metals?

FIGURE 1.6.1c: Credit cards contain a magnetic strip which encodes information about its owner.

Properties of non-metals

Non-metals have lower densities than metals, are often **dull** and are poor conductors of both heat and electricity. Around half are unreactive gases at room temperature. They have different uses to metals.

One non-metal, bromine, is unusual because it is a liquid at room temperature. It is harmful if its vapour is breathed in but it can be used to treat swimming pool water and in pesticides.

FIGURE 1.6.1d: Both bromine and chlorine can be used to treat water in swimming pools.

Sulfur is a bright yellow solid. It has uses in making rubber car tyres and gunpowder.

Neon can be used in glowing lights and helium is used to fill balloons.

> 6. Describe the differences in properties of metals and non-metals.
> 7. Describe what is unusual about bromine. Suggest why it is used in swimming pools.
> 8. Identify the property of helium that makes it useful in balloons.

Know this vocabulary

metal
non-metal
metals and non-metals
conduct
ductile
malleable
sonorous
magnetic
alloy
dull

SEARCH: properties of metals and non-metals

Reactions

Exploring the reactions of metals with acids

We are learning how to:
- Describe the reaction between acids and metals using word equations and particle diagrams.
- Explain the reaction between acids and metals.
- Compare the reactivities of different metals.

Most metals react with acids. The way that a metal reacts varies, depending on its reactivity. Some metals are so reactive that we would never add acid to them in the laboratory.

Reacting acids with metals

A **chemical reaction** is a change in which new products are made. There are clues that we can look for to spot a chemical reaction. These include:

- bubbles of gas being given off;
- a change in temperature;
- a colour change;
- a change in mass.

When we add an **acid** to most metals, we see bubbles. This is because a gas is produced during the reaction. We may also feel the test tube getting warmer. These observations are both evidence that a chemical reaction has taken place.

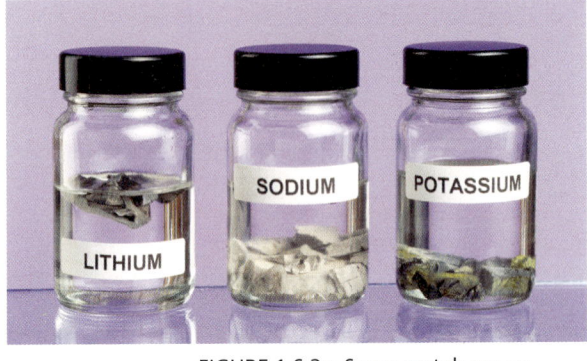

FIGURE 1.6.2a: Some metals are so reactive that they are stored under oil.

> 1. Describe some of the observations that tell us that a chemical reaction is taking place.
> 2. Describe two signs that the reaction between an acid and a metal is a chemical reaction.
> 3. Explain why bubbles are produced during some reactions.

FIGURE 1.6.2b: How can you tell that a chemical reaction is taking place between the acid and the magnesium?

What are the new products?

Acids react with most metals. Particles rearrange and a **salt** and **hydrogen** gas are formed. You can test for hydrogen gas because it burns with a 'pop'. If you put a lighted splint into the top of the test tube in which an acid and a metal are reacting, you will hear a 'pop' sound. This is because the flame ignites the hydrogen and it explodes.

130   AQA KS3 Science Student Book Part 1: Reactions – Metals and non-metals *and* Acids and alkalis

We can summarise the reaction between an acid and a metal using an equation:

acid + metal → salt + hydrogen

The type of salt produced depends on the type of acid and the metal used. For example, if you react nitric acid with zinc metal, zinc nitrate is the salt formed:

nitric acid + zinc → zinc nitrate + hydrogen

We can also show the reaction using a particle diagram:

nitric acid + zinc → zinc nitrate + hydrogen

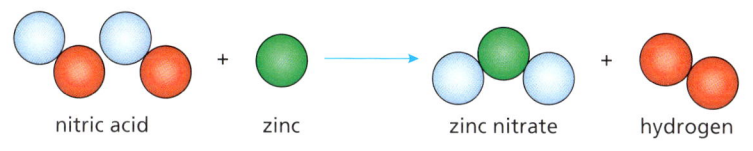

FIGURE 1.6.2d: Particle diagram for the reaction between nitric acid and zinc.

FIGURE 1.6.2c: Zinc nitrate crystals.

Did you know…?

Precious metals such as gold, silver and platinum do not react with acids. They are so unreactive that they stay as pure metals. This is one reason that they are used to make jewellery.

4. Write a word equation for the reaction between hydrochloric acid and magnesium metal.
5. Use the particle diagram in Figure 1.6.2d to explain how hydrogen gas is formed in the reaction between nitric acid and zinc metal.
6. Explain why we should not put a flame near a large amount of hydrogen gas.

Comparing reactivity

A group of students reacted hydrochloric acid with some different metals. They recorded their observations about the **reactivity** of the acid with the metals.

Metal	Observations when acid added
zinc	some bubbles
magnesium	vigorous bubbling
iron	a few bubbles
copper	no bubbles

FIGURE 1.6.2e: Results of the students' experiment showing reactions between hydrochloric acid and some different metals.

7. Order the metals in Figure 1.6.2e in terms of reactivity, going from most reactive to least reactive.
8. The teacher told the students that calcium is more reactive than the metals used in this investigation. Suggest what might be seen if the same acid was added to calcium.
9. Write a word equation for each of the reactions in Figure 1.6.2e.

FIGURE 1.6.2f: Precious metals are unreactive.

Know this vocabulary

chemical reaction
salt
hydrogen
reactivity

SEARCH: reactions of metals with acids

Reactions

Understanding displacement reactions

We are learning how to:
- Represent and explain displacement reactions using equations and particle diagrams.
- Make inferences about reactivity from displacement reactions.

We can use the order of reactivity of substances to make predictions about reactions. Reactive metals can be thought of as 'chemical bullies'. Why might this be so?

Chemical bullies

When a reactive metal reacts with a compound of a less reactive metal, the more reactive metal 'pushes out' or 'displaces' the less reactive metal. The more reactive metal forms a chemical bond with whatever the less reactive metal was bonded to. This can be shown using a particle diagram.

The situation is a bit like a basketball match. Imagine a weak player with the ball. A stronger player takes the ball from him, displacing the weaker player and leaving him on his own. An example of such a **displacement reaction** is when iron is added to a blue copper sulfate solution. Iron is more reactive than copper. A chemical change occurs – iron displaces the copper, bonding with the sulfate to make iron sulfate, which is a pale green solution.

The word equation for the reaction is:

iron + copper sulfate → iron sulfate + copper

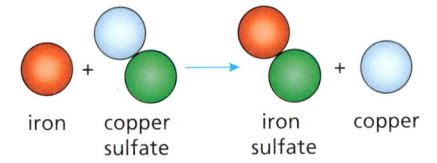

FIGURE 1.6.3a: Particle diagrams for the displacement of copper by iron.

FIGURE 1.6.3b: Iron and copper sulfate solution – before and after the displacement reaction. Over time, the blue copper sulfate solution becomes paler, and the iron nail becomes covered with a brown coating of copper.

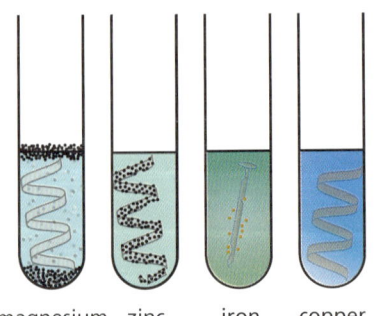

1. Describe how you know that a chemical reaction has taken place in Figure 1.6.3b.
2. When magnesium is added to a solution of copper sulfate, the solution changes from blue to colourless much faster than with iron. Which is more reactive, magnesium or iron?

Using displacement reactions

We can use displacement reactions to compare the reactivity of metals. We can produce a **reactivity series** (an order of reactivity) for magnesium, copper, iron and zinc, by adding each metal in turn to solutions of salts of each metal, for example, solutions of magnesium sulfate, copper sulfate, iron sulfate and zinc sulfate. Figure 1.6.3d shows the results that some students obtained after carrying out this experiment.

FIGURE 1.6.3c: Working out a reactivity series by reacting metals with copper sulfate solution.

6.3

metal \ metal sulfate solution	magnesium sulfate	zinc sulfate	iron sulfate	copper sulfate
zinc	✗		✓	✓
magnesium		✓	✓	✓
iron	✗	✗		✓
copper	✗	✗	✗	

Key: ✓ = reaction observed   ✗ = no reaction observed

FIGURE 1.6.3d: Results of students' experiment.

How might these students have decided whether or not there had been a reaction? Is there anything that they could have measured to show that a reaction had taken place?

3. A student couldn't identify one of the metals. It reacted with iron sulfate and copper sulfate but not magnesium sulfate. Which metal was it?
4. Write a word equation for the reaction between zinc and iron sulfate.
5. Using the results (Figure 1.6.3d), suggest a reactivity series (from most reactive to least reactive) for magnesium, copper, iron and zinc.
6. Other students included lead and lead nitrate solution in their investigation. They concluded that lead is more reactive than copper but less reactive than iron. Describe what they would have recorded in their results table for the reactions between:
   a) lead and copper sulfate;
   b) lead and iron sulfate.

Did you know...?

Old copper mines often become flooded, and a blue solution of copper sulfate results. By adding cheap scrap iron to this solution, copper metal is produced. This makes extra money for the mine owners.

Making predictions

The reactivity series is shown in Figure 1.6.3e. The further substances are from each other in the series, the more vigorous the displacement reaction between the more reactive substance and a salt of the less reactive substance.

The reactivity series consists mainly of metals. However, the non-metals carbon and hydrogen are often included because they can be used to extract metals that come below them in the series.

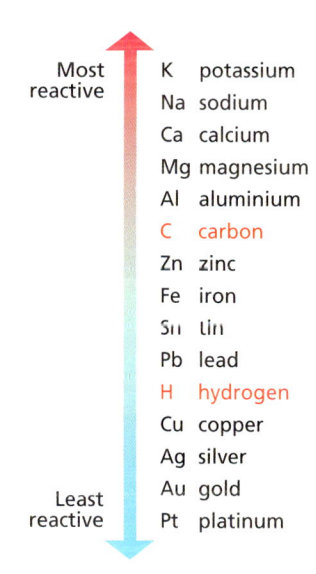

Most reactive
K   potassium
Na sodium
Ca calcium
Mg magnesium
Al aluminium
C   carbon
Zn zinc
Fe iron
Sn tin
Pb lead
H   hydrogen
Cu copper
Ag silver
Au gold
Least reactive
Pt platinum

FIGURE 1.6.3e: The reactivity series.

7. Why is no hydrogen produced when copper is added to hydrochloric acid?
8. Deduce a rule about how to use the reactivity series to predict whether a reaction will take place or not.

Know this vocabulary

displacement reaction
reactivity series

SEARCH: displacement reactions   133

Reactions

Understanding oxidation reactions

We are learning how to:
- Recall examples of oxidation reactions.
- Describe oxidation using word equations and particle diagrams.
- Investigate changes caused by oxidation.

Oxidation is an important chemical reaction that causes big changes. Some oxidation reactions are fast and others are slow. Oxidation is sometimes useful and at other times it causes problems. Browning of apples, rusting and burning are all oxidation reactions.

Oxidation

Oxidation is the name given to a chemical reaction in which oxygen is added to a substance. When a metal such as copper is heated in air it reacts with oxygen. Black copper oxide is formed:

copper + oxygen → copper oxide

We can also show these reactions using particle diagrams:

copper   oxygen   copper oxide

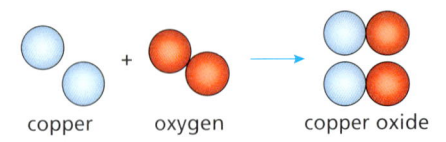

FIGURE 1.6.4a: Particle diagram for the reaction between copper and oxygen.

1. What is an oxidation reaction?
2. Give three everyday examples of oxidation reactions.
3. What changes would you see as copper is heated in the Bunsen burner flame?

Did you know...?

In 1986 a huge explosion occurred on the space shuttle *Challenger*, killing all seven astronauts aboard. This was caused by pure oxygen and hydrogen leaking into the shuttle's flames, leading to a powerful, uncontrolled oxidation reaction.

FIGURE 1.6.4b: *Challenger* orbiting the Earth.

Examples of oxidation

Iron and steel (an alloy of iron) undergo rusting. Rusting is an oxidation reaction where the metal reacts with oxygen in the air (when water is also present) to form iron oxide, or 'rust'. Stainless steel is an alloy of iron that doesn't rust. Every year, a huge amount of money is spent on preventing rusting of iron on buildings and structures, for example by painting or otherwise coating the metal.

6.4

There are many examples of useful oxidation reactions. **Combustion** is a special example of oxidation. Fuels contain the element carbon. During combustion, the carbon reacts with oxygen to form carbon dioxide gas. This gas escapes into the atmosphere.

Rockets use an oxidation reaction to fuel them. Hydrogen gas and pure oxygen gas are combined; the hydrogen is oxidised to water and an enormous amount of energy is given out.

Oxidation reactions can be used to make acids and bases (a **base** is any substance that neutralises an acid). Generally, non-metal oxides are acids (for example, sulfur dioxide) whereas metal oxides are bases (for example, sodium oxide).

FIGURE 1.6.4c: What is being oxidised in a combustion reaction?

4. Write a word equation for the formation of rust.
5. How can you tell that a chemical change has taken place during the oxidation reactions of rusting and combustion?

Investigating changes during oxidation

Two students carried out an experiment to investigate the oxidation of magnesium. They measured the mass of magnesium before heating it in a small container called a crucible (see Figure 1.6.4d). They observed a change in the magnesium from a silvery metal to a white powder. After the reaction was complete, the mass of the crucible was measured again.

The results of the experiment are shown in Table 1.6.4.

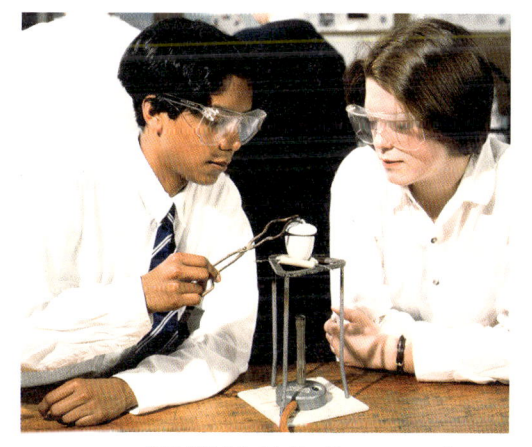

FIGURE 1.6.4d: Heating magnesium, allowing air in but no magnesium oxide to escape.

TABLE 1.6.4: Experiment results.

	At start	After heating	Change in mass
Mass of crucible and magnesium (g)	17.52	17.82	+0.30

6. Explain the changes that have taken place during the experiment.
7. Write a word equation for the reaction.
8. Using circles to represent the atoms, draw a particle diagram to explain the change in mass.

Know this vocabulary

oxidation
combustion
base

SEARCH: oxidation reactions   135

Reactions

Exploring acids

We are learning how to:
- Describe what an acid is and give examples.
- Identify the hazards that acids pose.

Acids are often thought of as dangerous substances. Indeed, many acids *are* dangerous and we must take precautions when handling them. However, we come across many acids in our daily life that are useful and not dangerous at all.

Useful acids

If you look around your kitchen, you may find some **acids** to eat or drink. Citrus fruits such as lemons and oranges contain citric acid. Vinegar, which is used to pickle foods or to flavour chips, contains ethanoic acid (sometimes called acetic acid). Fizzy drinks contain carbonic acid. Tea contains tannic acid. These acids tend to taste sour.

Acids also have industrial uses. Sulfuric acid is used in car batteries and in making fertilisers. Nitric acid can also be used in making fertilisers and in paints.

1. List some examples of acids that we have in our homes.
2. Describe two acids that may be used to make fertilisers.

Considering the hazards

Some acids, such as concentrated sulfuric acid, are extremely dangerous. These acids are **corrosive** – this means that the acid can destroy skin and attack metals if spilled.

The types of acid that are used in science lessons are dilute acids – this means that they have fewer acid particles than a concentrated acid solution in the same volume (Figure 1.6.5b). Dilute acids are not as dangerous as concentrated acids. They are not corrosive but may be an **irritant** to the skin. Your skin might become red and blistered if some laboratory acid were spilled on it.

Acids that are found in food and drink, such as in lemons and vinegar, are extremely weak and dilute. This is why they are safe to eat and drink, whereas dilute hydrochloric acid is not. However, they may still sting if they get into a cut.

FIGURE 1.6.5a: Which acid is found in each of these?

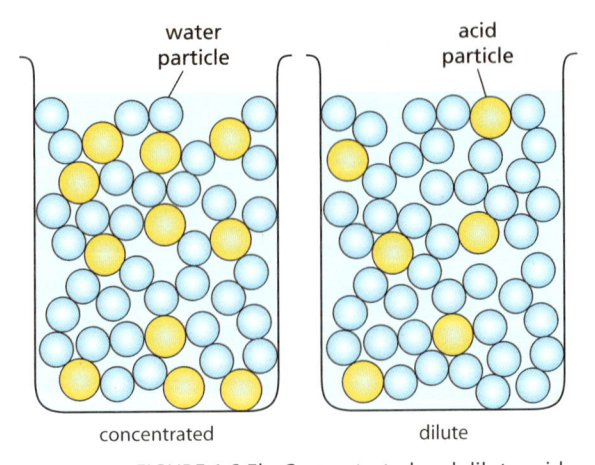

FIGURE 1.6.5b: Concentrated and dilute acids.

136 AQA KS3 Science Student Book Part 1: Reactions – Metals and non-metals *and* Acids and alkalis

3. Explain why it is better to use images on hazard labels, rather than words.
4. Describe the precautions that you should take when working with an acid that displays the 'warning' or 'irritant' hazard symbol.
5. Explain why concentrated acids are more dangerous than dilute acids.

What do acids have in common?

Some acids are weak enough that we can eat or drink them. Acetic acid and citric acid are weak acids. Some acids are strong and could burn your skin even when dilute. Hydrochloric, sulfuric and nitric acid are strong acids. One thing that all acids have in common is that they contain **hydrogen**, but not all compounds containing hydrogen are acids. We can show this by looking at the chemical formulas of acids:

Hydrochloric acid, HCl – this shows that the acid contains hydrogen (H) and chlorine (Cl).

Sulfuric acid, H_2SO_4 – this shows that the acid contains hydrogen (H), sulfur (S) and oxygen (O).

We measure the strength of an acid using a scale called the **pH** scale. The scale ranges from 1 to 6 for acids. The stronger the acid, the lower its pH. We can use strong acids in the laboratory without too much danger if we use them at a low enough **concentration**.

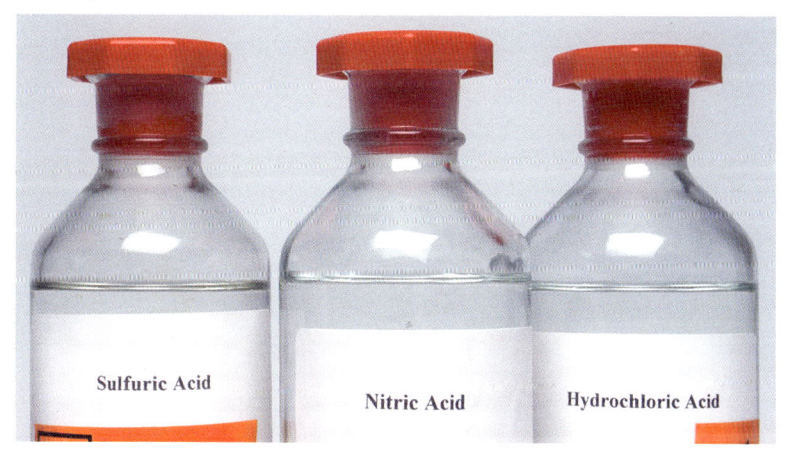

FIGURE 1.6.5c: All of these acids contain hydrogen.

6. The chemical formula for nitric acid is HNO_3. Which elements does nitric acid contain?
7. A sour-tasting substance is found to contain the elements oxygen, sulfur and hydrogen. Suggest whether or not this is an acid and explain your reasoning.

Did you know...?

Your stomach contains hydrochloric acid, which helps to digest food and kill bacteria. You can feel this acid burning your throat slightly when you vomit.

FIGURE 1.6.5c: 'Corrosive' hazard sign.

FIGURE 1.6.5d: Irritant hazard sign, which is used for substances that are not corrosive but are irritants.

Know this vocabulary

acid
corrosive
irritant
hydrogen
pH
concentration

SEARCH: uses of acids   137

Reactions

Exploring alkalis

We are learning how to:
- Describe what an alkali is and give examples.
- Identify the hazards that alkalis pose.

Many of the cleaning products that we use have something in common – they all contain an alkali. It is the alkali that gives soap, shampoo and washing powder a soapy feeling. We have alkalis all around us and life would be very different without them.

Useful alkalis

Some **alkalis** are harmful. However, many alkalis are harmless and are very useful.

Many cleaning products – such as bleach, oven cleaner, disinfectant and washing powder – contain alkalis. Toiletries such as soap, shampoo and toothpaste also contain an alkali.

Indigestion remedies contain alkaline substances.

When you bake a cake, you use baking powder to ensure that the cake is light and fluffy. Baking powder contains an alkali called sodium hydrogencarbonate (sodium bicarbonate). Without it, your cakes would be like biscuits!

1. Name some alkaline cleaning products.
2. Name two alkaline substances that are safe to put in your mouth and two that are not.
3. Suggest how your life would change if there were no alkalis.

FIGURE 1.6.6a: Many cleaning products contain an alkali.

Considering the hazards

Many of the alkalis in our homes are dangerous. The most dangerous alkalis include oven cleaners and caustic soda (to unblock drains). These substances are corrosive – they both contain the alkali sodium **hydroxide**.

Other alkalis are classed as an irritant, rather than corrosive. Examples are bleach and disinfectant.

Alkalis are often more dangerous than acids given the same hazard classification. This is because it can be hard to rinse an alkali from the skin because it becomes soapy.

6.6

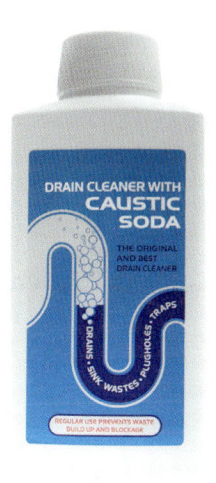

FIGURE 1.6.6b: Which alkali do both of these products contain?

Did you know…?

In the past, stale urine was used as a source of the alkali ammonium hydroxide. It was used to bleach and clean clothes – it was even used in toothpaste!

4. Bleach used in homes often has a warning written in Braille on the bottle. Suggest why this is important.
5. Draw the hazard symbol that would be found on a bottle of bleach.
6. Bleach contains sodium hydroxide and another chemical, sodium hypochlorite. Bleach is dangerous, but caustic soda is even more dangerous. Suggest why.

What do alkalis have in common?

Most alkalis feel soapy to touch. Soap forms because the alkali reacts with fats on your skin. However, some alkalis are too harmful to put on your skin. The common feature of all alkalis is that they contain hydroxide particles (chemical symbol OH).

Sodium hydroxide, NaOH, is the alkali used in many cleaning products, such as oven cleaners. Calcium hydroxide, $Ca(OH)_2$, is an alkali used by gardeners when their soil is too acidic. Both of these products would be harmful if you swallowed them. Magnesium hydroxide is the weak alkali found in some indigestion remedies.

FIGURE 1.6.6c: Which alkali does baking powder contain?

We can measure the strength of an alkali using the pH scale. The scale ranges from 8 to 14 for alkalis. The stronger the alkali, the higher its pH.

7. What is the common feature of all alkalis?
8. Which elements are contained in:
    a) calcium hydroxide?   b) sodium hydroxide?
9. If alkali A has a pH value of 9 and alkali B has a value of 12, which is the stronger alkali? Which is most likely to be used in indigestion remedies?

Know this vocabulary

alkali
hydroxide

SEARCH: alkalis in the home   139

Reactions

Using indicators

We are learning how to:
- Use indicators to identify acids and alkalis.
- Analyse data from different indicators.
- Compare the effectiveness of different indicators.
- Describe what a pH scale measures.

The traffic indicators on a car tell other vehicles when the car is going to turn. Indicators in science can show us whether a substance is an acid or an alkali. Nature is full of natural indicators and we can make use of these indicators in many ways.

What are indicators?

An **indicator** is a substance that has different colours in an acid and in an alkali. One example of an indicator is **litmus**. Litmus solution turns *red in acid* and *blue in alkali*. If a solution is neither an acid nor an alkali, we say it is **neutral**.

Litmus paper is sometimes easier to use than litmus solution. Blue litmus paper turns red in an acid; red litmus paper turns blue in an alkali.

FIGURE 1.6.7a: What colour is litmus in an acid and in an alkali solution?

1. Describe what an indicator is.
2. Describe the colour changes of litmus solution in an acid and an alkali.
3. Draw a table to show the colours in acid, alkali and neutral of:
   a) red litmus paper;   b) blue litmus paper.

Using universal indicator

Most chemical indicators just tell us whether a substance is an acid or an alkali. **Universal indicator** turns a range of different colours. The colour depends on whether the substance is an acid or an alkali *and* on how strong or weak it is (Figure 1.6.7b). Each colour is given a **pH** number.

On the pH scale:
- neutral solutions are pH 7;
- acidic solutions are lower than pH 7;
- alkaline solutions are higher than pH 7.

Did you know…?

Baking powder can also be used as an indicator. It does not show any colour change but it does fizz when added to an acid, but not when added to an alkali or to water.

6.7

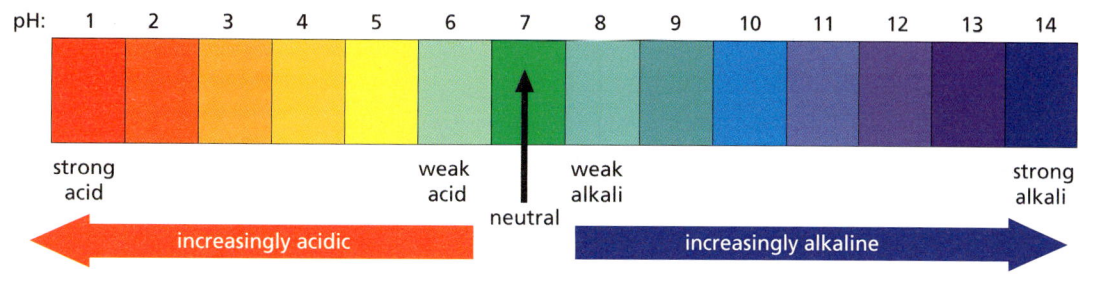

FIGURE 1.6.7b: The colour of universal indicator shows the strength of acids and alkalis. The pH scale ranges from pH 1 to pH 14.

4. Universal indicator is added to a liquid and it changes to yellow. State the pH of the liquid.
5. Describe what happens to the strength of an acid as the pH number decreases.
6. Describe what happens to the strength of an alkali as the pH number increases.

Comparing indicators

Litmus indicator turns red in acid and blue in alkali. Red cabbage indicator turns red in acid and purple in alkali.

Universal indicator is a mixture of several different indicators. This means that it gives a full range of predictable colours, depending on the strength of the acid or alkali.

TABLE 1.6.7: The pH values of different acids and alkalis.

Substance	pH	Acidic or alkaline?
hydrochloric, nitric and sulfuric acids, and car battery acid	0–1	strongly acidic
phosphoric acid	1–2	acidic
citrus fruit, such as lemons and oranges; vinegar	4	acidic
distilled water	7	neutral
egg, hand soap	8	alkaline
ammonia	11	alkaline
oven cleaner	12	alkaline
caustic soda, paint stripper	13–14	strongly alkaline

7. Describe what colour litmus indicator would turn if added to:
   a) hydrochloric acid;
   b) vinegar.
8. Explain the advantages of using universal indicator over litmus or red cabbage indicator.

Know this vocabulary

indicator
litmus
neutral
universal indicator
pH

SEARCH: universal indicator 141

Reactions

Exploring neutralisation

We are learning how to:
- Recall and use the neutralisation equation.
- Use indicators to identify chemical reactions.
- Explain colour changes in terms of pH and neutralisation.

The pain of a nettle sting can be eased by rubbing the sting with a dock leaf. Nettles contain a weak acid and dock leaves contain a weak alkali. The alkali 'cancels out' the acid. This is called neutralisation, and there are many other examples around us.

FIGURE 1.6.8a: Why does a dock leaf help with a nettle sting?

Mixing acids and alkalis

As we add an alkali to acid, the particles in the acid and alkali react. The resulting solution becomes less acidic (the pH increases) as we add more alkali. This reaction between acids and alkalis is called **neutralisation**.

If we add just the right amount of alkali, the solution will become exactly neutral.

1. Describe what is meant by 'neutralisation'.
2. Describe what happens to the pH of an acidic solution as an alkali is added. Explain your answer.

Demonstrating neutralisation

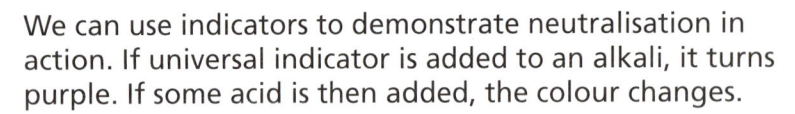

We can use indicators to demonstrate neutralisation in action. If universal indicator is added to an alkali, it turns purple. If some acid is then added, the colour changes.

We can use a technique called **titration** to mix acids and alkalis precisely. This allows us to see a whole range of colour changes.

A burette allows an acid to be added to an alkali gradually. If the acid is added slowly enough, the neutral point (pH 7) can be seen. This point is indicated by the solution turning green.

3. Describe the colour changes that would be seen in the conical flask as the solution changed from a strong alkali to neutral.

Did you know…?

Bee stings are acidic and are treated by neutralising the acid with a mild alkali, such as bicarbonate of soda. Wasp stings are slightly alkaline and are treated by neutralising with an acid, such as vinegar. Therefore, it is important to know what has stung you.

4. Suggest what would be seen if more acid were added after the neutral point was reached. Explain your answer.
5. Explain the benefits of using a burette, rather than dropping pipettes, to add the acid.

The neutralisation equation

We can describe neutralisation using an equation:

acid + alkali → salt + water

If hydrochloric acid is neutralised with the alkali sodium hydroxide, the salt produced is sodium chloride.

The first part of the name of the salt comes from the alkali, usually from the metal in the alkali. For example, the alkali sodium hydroxide forms salts that start with 'sodium', whereas magnesium hydroxide forms salts that start with 'magnesium'.

The second part of the name of the salt comes from the acid. Table 1.6.8 summarises the ends of the salt names for each of the common acids.

TABLE 1.6.8: The acid used tells us the end of the salt name.

Acid used in neutralisation	Forms salts that end in...
hydrochloric acid	chloride
sulfuric acid	sulfate
nitric acid	nitrate

6. Write the general equation for neutralisation.
7. Name the product of neutralisation that:
    a) is always the same;
    b) depends on the acid and alkali used.
8. a) Write an equation for the reaction between hydrochloric acid and sodium hydroxide.
   b) Describe the two new products that are formed when hydrochloric acid is neutralised with sodium hydroxide.

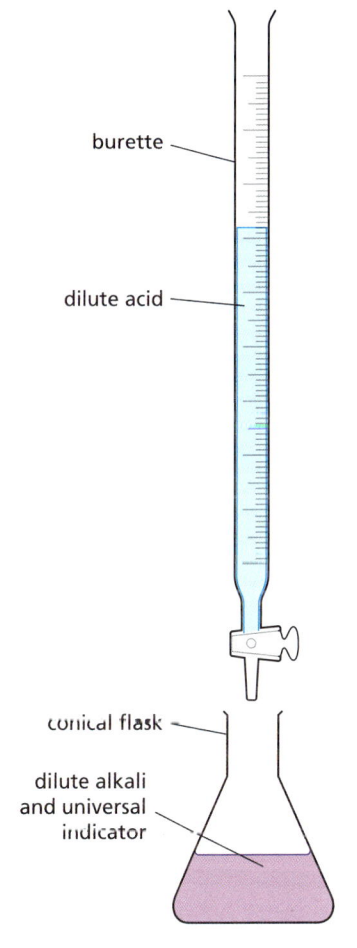

FIGURE 1.6.8b: Titration can be used to carry out a neutralisation reaction precisely.

Know this vocabulary

neutralisation
titration

SEARCH: neutralisation

Reactions

Investigating neutralisation

We are learning how to:
- Design an investigation to compare the effectiveness of indigestion remedies.
- Analyse data to identify a suitable indigestion remedy and suggest improvements to the investigation.

Heartburn indigestion is caused by acid from the stomach irritating the upper digestive tract. For those who suffer regularly, treatments are available to neutralise this acid. But are some remedies more effective than others?

The need for antacids

The human stomach contains strong hydrochloric acid, with a pH of approximately 1. The acid helps enzymes to digest proteins in the stomach and also prevents some bacteria from surviving in the stomach. Heartburn is a type of indigestion caused when the muscle leading from the osesophagus to the stomach opens, allowing stomach acid to move up the digestive tract. The acid causes a burning sensation in the chest.

Medicines called **antacids** contain substances that can neutralise the acid from the stomach. These substances are bases (remember any substance that neutralises an acid is a **base**, and alkalis are soluble bases). Examples of these bases are calcium carbonate, magnesium hydroxide and sodium hydrogencarbonate (baking soda).

$$acid + base \rightarrow salt + water$$

1. Explain what heartburn is and what causes it.
2. Explain how antacids reduce the acidity of stomach acid.
3. Suggest the effect of antacid remedies on the pH of stomach acid.

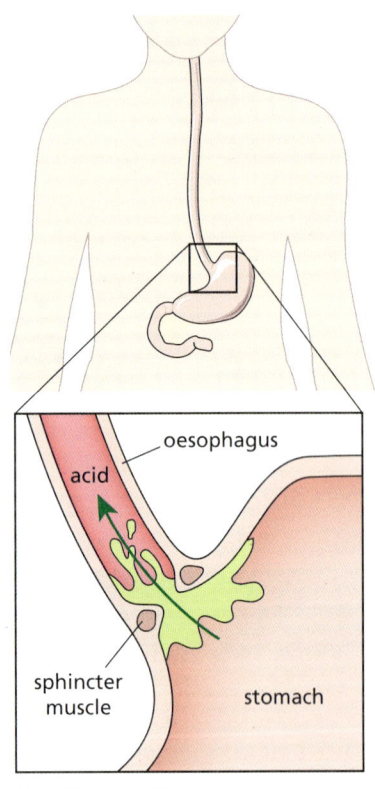

FIGURE 1.6.9a: What causes heartburn?

Planning an investigation

A group of students wanted to compare the effectiveness of different commercial antacid remedies.

When we plan an investigation, we must consider **variables**. Things we could change are called *independent* variables. Altering these could make a difference to other things, which are called *dependent* variables. Some of the independent variables will be kept the same; these are then called *control* variables. If we alter one independent variable and see how a dependent variable changes we can look for a **correlation**.

TumCalm®
Indigestion relief

Ingredients
Calcium carbonate
Magnesium hydroxide
Gelling agent
Water
Peppermint flavouring
Permitted sweetener
Colouring

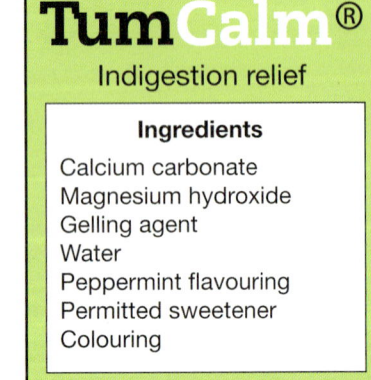

FIGURE 1.6.9b: Which bases are found in this remedy?

In this investigation, the students are changing the type of indigestion remedy. They are going to measure the time taken for the pH to change from 1 to 7.

6.9

4. Identify the independent variables and the dependent variables in this investigation.

5. Suggest which variables the students should control during this investigation.

6. The students devised a question for their enquiry, 'Which indigestion remedy is best?'. Suggest an improvement on this question using the stem, 'How does...affect...?'

Did you know...?

Even before the chemistry was understood, acid-neutralising remedies were recommended for heartburn. One remedy was to chew limestone rock. We now know that limestone contains the base calcium carbonate.

Considering the data

The students added universal indicator to a beaker of hydrochloric acid at pH 1. They added the recommended dose of indigestion remedy to the acid and measured the time taken for the pH to change from pH 1 to pH 7. They repeated this for each indigestion remedy twice.

Following an investigation, it is important that we consider whether we can be sure that we can trust our data. If the repeat readings are close together, that suggests that our experiment is **repeatable**. If this is the case, we can calculate a **mean** value from our data. If our repeat readings are not close together, we can choose to ignore a reading when calculating a mean or we could repeat that part of the investigation.

Leading from this, we could suggest improvements to our investigations, for example, should we repeat any particular readings, should we have carried out an additional set of readings, or could we have used more accurate measuring equipment?

Antacid remedy	Time taken (s)	
	1	2
Acid-ban	360	388
Acid-ease	175	192
Banish burn	556	544

TABLE 1.6.9: Results of tests on antacid remedies.

7. The students' results are shown in Table 1.6.9. Suggest whether this experiment was repeatable and explain your answer.

8. a) Calculate a mean value of the time taken for each indigestion remedy.

   b) Write a conclusion about which remedy is most effective. Provide scientific explanations for your conclusion.

9. Suggest improvements that these students could make to their method, to improve the accuracy of their results.

Know this vocabulary

antacid
base
variable
correlation
repeatable
mean

SEARCH: examples of neutralisation

Reactions

Checking your progress

To make good progress in understanding science you need to focus on these ideas and skills.

☐ Identify some common properties of metal elements and non-metal elements and their uses.	☐ Classify metals and non-metals using their properties.	☐ Identify similarities and differences between metals and how these relate to their uses; compare and contrast properties of metals and non-metals.
☐ Identify oxidation reactions.	☐ Explain why oxidation is a reaction.	☐ Use models and word equations to explain changes during oxidation reactions.
☐ Give uses of displacement reactions.	☐ Use models to explain displacement and relate it to the reactivity series.	☐ Write word equations to represent displacement reactions.
☐ Identify some everyday substances that contain acids and alkalis.	☐ Explain what all acids have in common and what all alkalis have in common.	☐ Evaluate the hazards posed by some acids and alkalis and know how these risks may be reduced.
☐ Give an example of an indicator and state why indicators are useful.	☐ Explain what an indicator is and analyse results when using an indicator.	☐ Compare the effectiveness of different indicators.

6.10

- ☐ Describe some examples of neutralisation.
- ☐ Describe the changes to indicators when acids and alkalis are mixed.
- ☐ Explain the changes to indicators in terms of pH when acids and alkalis are mixed.

- ☐ Recognise that water is one product of neutralisation.
- ☐ Explain the formation of salt and water during neutralisation, giving some examples of common salts.
- ☐ Predict the reactants or products of different neutralisation reactions.

- ☐ Describe what indigestion remedies are and explain how they work.
- ☐ Design an investigation to compare the effectiveness of indigestion remedies.
- ☐ Analyse data about indigestion remedies to decide which remedy is the most effective.

Reactions

Questions

KNOW. Questions 1–9

See how well you have understood the ideas in this chapter.

1. What name is given to a reaction in which a chemical combines with oxygen? [1]

    a) aeration   **b)** oxygenation   **c)** oxidation   **d)** breathing

2. Write a word equation for the reaction between zinc and hydrochloric acid. [2]

3. Give two examples of the differences between the properties of metals and non-metals. [2]

4. When magnesium is heated, it changes from a silver colour to a white powder. Is this a physical process or a chemical reaction? Explain your answer. [2]

5. All acids contain the element: [1]

    a) hydrogen   **b)** oxygen   **c)** chlorine   **d)** hydroxide

6. What is the pH of a neutral solution? [1]

    a) 1   **b)** 14   **c)** 7   **d)** 10

7. The reaction between an acid and an alkali is known as: [1]

    a) neutralisation   **b)** oxidation   **c)** burning   **d)** combustion

8. Write a word equation for the reaction between an acid and an alkali. [2]

9. Explain why heartburn is treated using a base. [2]

APPLY. Questions 10–14

See how well you can apply the ideas in this chapter to new situations.

10. What product is formed when iron and oxygen react together? [1]

    a) oxyiron   **b)** iron oxygen   **c)** iron oxate   **d)** iron oxide

11. Magnesium reacts with oxygen to form magnesium oxide. The mass of the magnesium oxide at the end of the experiment is greater than the mass of magnesium at the start because: [1]

    a) it burns with a bright light;
    b) it gives off carbon dioxide;
    c) magnesium is not very dense;
    d) the oxygen has added to the mass of the magnesium.

12. Calcium carbonate can be used to relieve indigestion because: [1]

    a) it digests food;   **b)** it tastes nice;
    c) it increases acidity;   **d)** it reduces acidity.

6.11

13. A chemical is described as feeling 'soapy'. When tested with indicator, it is shown to have a pH of 9. Explain what type of chemical this is. [2]

14. A student notices that a concentrated acid gave a more vigorous reaction with a metal than a dilute acid with the same metal. Explain why, using the idea of particles. [2]

EXTEND. Questions 15–17

See how well you can understand and explain new ideas and evidence.

15. A student reacts different metals with hydrochloric acid. The observations are recorded in Table 1.6.11a. One of the metals is not labelled.

    TABLE 1.6.11a

Metal	Observations
unknown metal	Bubbles seen, test tube became warmer
calcium	Lots of bubbles produced very quickly, test tube became very hot
zinc	Bubbles seen

    Compare the reactivity of the unknown metal with that of calcium and zinc. Explain your answer. [2]

    FIGURE 1.6.11a: Zinc metal reacting with hydrochloric acid.

16. Brass is a gold-coloured metal – it is an alloy of copper and zinc. It can be cast into different shapes and has a range of uses including musical instruments, electrical switches and door fittings. From this information suggest some of the properties of brass, explaining your answers. [4]

17. A group of students tried to make some indicators from different plant materials. They tested each of the solutions that they made. The results are shown in Table 1.6.11b.

    TABLE 1.6.11b

Indicator	Colour in acid	Colour in alkali	Colour in neutral
A	red	blue	blue
B	red	blue	purple
C	yellow	yellow	yellow

    Arrange the indicators in the order of most useful to least useful for testing the pH of a variety of different chemicals. Explain your answer. [3]

Earth
Earth structure *and* Universe

Ideas you have met before

Rocks have properties which can be studied
Rocks can be grouped together based on their appearance, such as whether they have grains or crystals.

Different kinds of rock can be compared and grouped together on the basis of their physical properties.

Formation of rocks
Fossils are formed when organisms are trapped within the layers of sedimentary rock.

Soils are made from rocks and organic matter.

The Earth in space
The Sun, the Earth and the Moon are approximately spherical objects.

The Sun is our nearest star but there is an unimaginable number of other stars.

The Earth and other objects in space move
The movement of the Earth, and other planets in the solar system, can be described relative to the Sun.

The Earth's daily spinning motion explains day and night and the apparent movement of the Sun across the daytime sky.

The movement of the Moon can be described relative to the Earth.

AQA KS3 Science Student Book Part 1: Earth – Earth structure *and* Universe

7.0 In this chapter you will find out

The rock cycle
- Sedimentary, igneous and metamorphic rocks can be inter-converted over millions of years, through weathering and erosion, heat and pressure, and melting and cooling.
- Magma from volcanoes solidifies to form igneous rock.
- There is a relationship between the shape of a volcano and the type of magma it produces.
- There are different ways that fossils can form in sedimentary rock.
- Rocks are continually being broken down and new rocks are formed. This is described by the rock cycle.
- The constant movement of the Earth's crust causes rocks deep underground to be brought to the surface and mountain ranges to form.

The Earth in the Universe
- Distances in space are so vast that special units are used to measure them.
- Our solar system is a tiny part of a galaxy, one of many billions of galaxies in the Universe.
- Light takes minutes to reach Earth from the Sun, four years from our nearest star and billions of years from other galaxies.

The movement of objects in space
- The solar system can be modelled as planets rotating on tilted axes while orbiting the Sun, moons orbiting planets, and sunlight spreading out and being reflected.
- This explains day and year length, the seasons, and how we see objects from Earth.

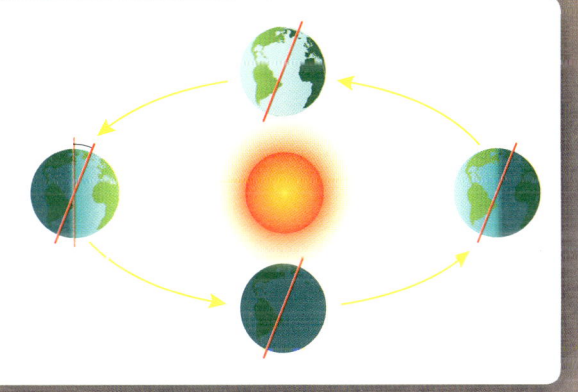

Earth

Understanding the structure of the Earth

We are learning how to:
- Name the layers of the Earth.
- Describe the characteristics of the different layers.
- Explain how volcanoes change the Earth.

The Earth has various layers, some of which are constantly moving. What are the different layers of the Earth called? What are their features?

The Earth's layers

The Earth is made up of different layers:
- **core** (part solid and part liquid);
- **mantle** (semi-liquid and solid);
- **crust** (solid).

The crust and the outer (solid) part of the mantle are called the **lithosphere**. This consists of pieces of rock called **tectonic plates** that float on the semi-liquid mantle and move about slowly.

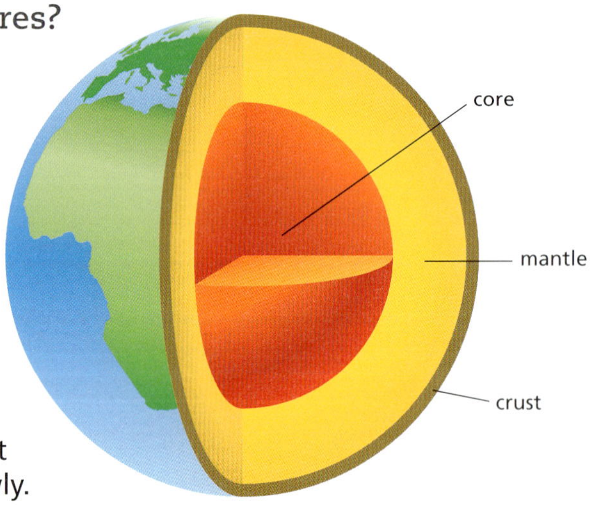

It is difficult to study the structure of the Earth directly because the crust is too thick to drill right through. However, scientists can study how waves made by earthquakes and explosions travel through the Earth. This gives them evidence of the different types of material in the different layers.

FIGURE 1.7.1a: The crust is the outer layer of the Earth – it is the land on which we live.

1. Suggest a kind of fruit which, when cut open, might look rather like a sectional view of the earth.
2. Why do scientists have earthquake sensors all over the world?

Features of the layers

The Earth's core is very hot. It consists of nickel and iron.

The mantle is the very thick middle layer (about 3000 km thick). It contains silicon, magnesium and iron, in the form of oxides. The flowing mantle material transfers heat outwards from the core.

The crust is relatively thin (5 km to 100 km thick) and rocky. There are two types – the dense, thinner oceanic crust (made of basalt) and the less dense continental crust (which is granite).

The Earth's lithospere is a relatively cold part of the Earth. It is made up of about 20 tectonic plates. These move at a rate of about 2.5 cm per year on average. Over millions of years this has allowed whole continents to shift thousands of kilometres apart. This process is called 'continental drift'.

7.1

3. Explain the difference between the two types of crust.
4. How do continents move?

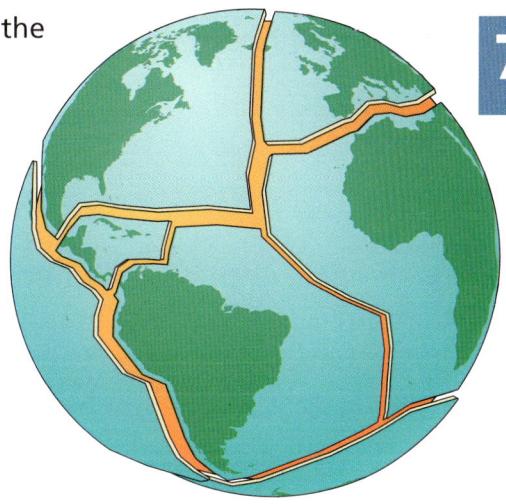

FIGURE 1.7.1b: The map of the Earth is changing very slowly because the plates are constantly moving.

Changing the Earth's surface

Where tectonic plates meet, they can push against each other, or move under or over each other. Earthquakes and volcanic eruptions occur at these points, and the crust may crumple to form mountain ranges. **Magma**, which is molten rock from the mantle, is less dense than the crust. It can rise to the surface through volcanoes (weak areas of the crust). **Lava** is the molten rock that escapes onto the Earth's surface. As this cools down it solidifies.

Geologists study volcanoes to try to predict future eruptions and to study the Earth's structure. Volcanoes can be very destructive. Even so, farming communities may choose to live near them because volcanic soil is very fertile.

Did you know...?

The cinder cone volcano Paricutin appeared in a Mexican cornfield on February 20, 1943. By the end of a year it was 336 m tall, and it reached its tallest height of 424 m in 1952. In geology, that is very quick

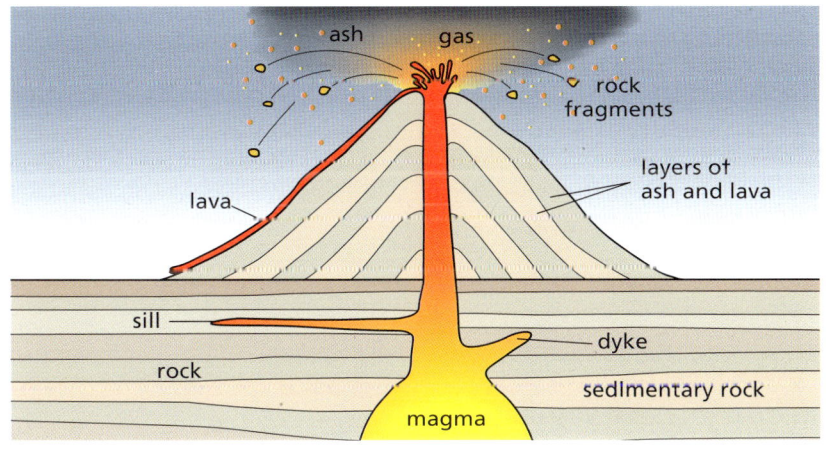

FIGURE 1.7.1c: The inside of a volcano, showing how the layers of ash and lava build up.

5. How do volcanoes form?
6. What is the difference between lava and magma?

Know this vocabulary

core
mantle
crust
lithosphere
tectonic plate
magma
lava

SEARCH: structure of the Earth 153

Earth

Exploring igneous rocks

We are learning how to:
- Describe how igneous rocks are formed.
- Explain how the pH of the magma affects the formation of rocks.
- Investigate the effect of cooling rate on the formation of crystals.

Some of the oldest rocks on Earth are igneous rocks. Other igneous rocks are being formed right now. The word 'igneous' comes from a Greek word for 'fire'. How do igneous rocks form? What are their features?

What are igneous rocks?

Igneous rocks form when hot molten rock from the Earth's mantle cools down and hardens. They have no layers, may be light- or dark-coloured, usually have crystals and rarely react with acids. They do not contain fossils because these would have melted when the magma formed.

There are two main types of igneous rock:

- **extrusive** – these form when magma flows onto the Earth's surface;
- **intrusive** – these form from magma below the Earth's surface in the crust.

Igneous rocks make up most of the rock on Earth, but they are often buried below the surface. One of the most common igneous rocks is granite, which is used for building and making statues. Other examples are pumice, basalt and obsidian.

FIGURE 1.7.2a: The Sierra Nevada mountains in the United States are made of granite.

1. Name four igneous rocks.
2. Describe some of the features of a typical igneous rock.

Looking at magma

Some volcanoes formed from acidic magma and volcanic ash. They are typically steep and conical, for example Mount Fuji in Japan. These volcanoes often exceed heights of 2500 m. They have periodic explosive eruptions. The acidic lava that flows from them is very viscous (thick and sticky). It cools and hardens before spreading very far. Rocks formed from acidic magma include granite, pegmatite and pumice.

7.2

FIGURE 1.7.2b: Mount Fuji was formed from acidic magma. Olympus Mons on Mars was formed from alkaline magma.

Other volcanoes formed from alkaline magma. They typically have shallow, sloping sides – for example, Olympus Mons on Mars and the Hawaiian volcanoes. They often eject large amounts of lava onto the ground. The alkaline lava that flows from them is thin and runny. It can travel long distances before it cools and hardens to form rocks. Rocks formed from alkaline magma include basalt and gabbro.

3. How does pH affect magma?
4. What is the relationship between magma viscosity and volcano shape?

Did you know…?

Obsidian forms from magma that cools so rapidly that no crystals develop – it forms a glass.

Crystal size

The rate at which lava or magma cools determines the size of the crystals in an igneous rock. If the rate of cooling is fast, the rock will have small crystals. If the rate of cooling is slow, the rock will have large crystals. Granite cools slowly and has large crystals; gabbro cools even more slowly and has even larger crystals. Intrusive rocks often cool more slowly than extrusive rocks. However, when **fissures** (cracks) open underground, the magma in them cools quickly to form rocks with small crystals (such as basalt).

FIGURE 1.7.2c: Look carefully at the magnified images of these two rocks. What conclusions can you come to about how they cooled?

5. What is the relationship between rate of cooling and crystal size?
6. Compare the rate of cooling in intrusive and extrusive rocks.

Know this vocabulary

igneous rocks
extrusive
intrusive
fissure

SEARCH: igneous rocks 155

Earth

Exploring sedimentary rocks

We are learning how to:
- Describe how sedimentary rocks are formed.
- Explain how fossils give clues about the past.
- Explain the properties of sedimentary rocks.

Sedimentary rocks are formed over thousands or even millions of years. What are their features? How are they formed?

Rocks in layers

Rocks suffer **weathering** and **erosion** – pieces break off and are then transported by wind or water. When river or sea currents slow down, rocks, pebbles and sediments drop to the riverbed or seabed – this is **deposition**. Over millions of years they are buried under more sediments. The weight of the upper layers compacts (presses together) and cements (sticks together) the lower sediments to form **sedimentary rocks**.

Sedimentary rocks are usually crumbly, found in layers called **strata** and can contain **fossils**. Examples are:
- sandstone – made of sand particles;
- limestone – made of tiny shells and skeletons of marine organisms;
- shale and mudstone – made of silt and clay particles that are too small to see;
- conglomerate – made of rounded pebbles.

1. How do rocks become sediments?
2. Name and describe three sedimentary rocks.

FIGURE 1.7.3a: Sedimentary rocks build up in layers and may contain fossils.

FIGURE 1.7.3b: Sedimentary rocks are made of rock particles and are usually porous, meaning water can pass through the gaps between the grains. This shows the grains in sandstone under a microscope.

Looking at fossils

A **fossil** is the preserved remains of a dead organism. Fossils give clues about the environment that the rock formed in. For example, they can tell us if it formed in fresh water or seawater.

Fossils form when dead organisms get covered in a layer of sediment before they can rot away. If the covering sediments change into sedimentary rocks, the remains of the animal or plant can also turn into rock but keep their original shape.

Did you know…?

Weathering is the *wearing down* of rock by physical, chemical or biological processes. Erosion is the *movement* of rock by water, ice or wind.

There are three main ways that fossils can form:
- hard body parts (shells or bones) can be covered by sediments and then replaced by **minerals**;
- softer parts of plants and animals can form casts or impressions;
- dead plants and animals can be preserved in amber (a sticky tree resin), tar pits or glaciers.

3. Is limestone made in the sea or on land? Explain your answer.
4. How do fossils form?

Breaking rocks

Rocks are gradually weathered – they wear away. Acid rain causes chemical weathering – it dissolves rocks such as marble and limestone. Waves pound on rocks and eventually cause cliffs to crumble. Fast water in rivers or strong waves on beaches pick up rocks, knocking off sharp edges and turning them into smooth weathered rock material. When the weathered rock material is deposited on the river or seabed, water seeps through the sediments. Minerals in the water can crystallise between the rock particles and cement them together.

Freeze–thaw weathering of rocks happens when water seeps into cracks in the rock and then freezes. As it freezes it expands, eventually breaking the rock apart.

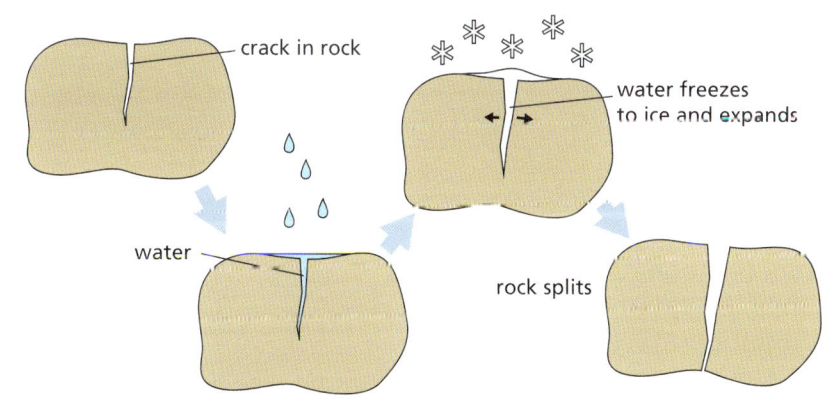

FIGURE 1.7.3d: How does water cause rocks to break apart?

Tree roots can also gradually break rocks apart as they grow. Some living organisms, such as bacteria and algae, produce chemicals that react with the rock and break it up.

5. What causes weathered rock material to become smooth?
6. Explain the processes of deposition, compaction and cementation.

FIGURE 1.7.3c: Fossils give us information. Dinosaur footprints can tell us about the dinosaur's size, weight and how it moved.

Did you know…?

Rocks and minerals are not the same thing. Minerals are the chemicals that rocks are made from.

Did you know…?

Sedimentary rocks cover most of the Earth's surface, but only make up a small percentage of the crust compared to metamorphic and igneous rocks.

Know this vocabulary

weathering
erosion
deposition
sedimentary rocks
strata
fossil
minerals
freeze–thaw

SEARCH: sedimentary rocks   157

Earth

Exploring metamorphic rocks

We are learning how to:
- Describe how metamorphic rocks are formed.
- Explain the properties of metamorphic rocks.

The word 'metamorphic' comes from the Greek for 'change of form'. What are metamorphic rocks changed from? What are their features?

Making metamorphic rocks

Existing rocks that are subjected to large amounts of heat and/or pressure can change into another type of rock called **metamorphic** rock. The original rocks are usually found deep in the Earth's crust. The new metamorphic rock is generally very hard-wearing and resistant to weathering and erosion.

Examples of metamorphic rocks are:
- marble formed from limestone;
- slate formed from clay;
- schists formed from sandstone or shale.

1. Describe how metamorphic rocks are formed.
2. What type of rocks are limestone, clay, sandstone and shale?

Metamorphic changes

When existing rocks **metamorphose**, they change their crystal structure without melting. New crystals form ('recrystallisation') and the structure of the original rock changes permanently. This can happen in and around volcanoes, for example. The new minerals are more stable in the new conditions of pressure and temperature.

Different minerals form at different temperatures. The new minerals can be used to estimate the temperature, depth and pressure that the original rock metamorphosed at.

Limestone, chalk and marble are chemically identical but only marble is a metamorphic rock. Metamorphic rocks are usually very hard and shiny. Marble is a typical example – it is extremely hard and can be polished. Marble is used by sculptors because it can be carved into complex shapes.

FIGURE 1.7.4a: What are the names of the metamorphic rocks used in these pictures?

Did you know…?

Metamorphic rocks can be formed from igneous, sedimentary or other metamorphic rocks, but the changes from sedimentary to metamorphic are the most extreme.

158 AQA KS3 Science Student Book Part 1: Earth – Earth structure *and* Universe

TABLE 1.7.4: Examples of metamorphic rocks and their uses.

Original rock	Metamorphic rock after metamorphism	Uses of the metamorphic rock
sandstone (sand grains in layers)	quartzite – much harder; original layers destroyed	building stone
limestone (layers, often with fossils)	marble – much harder; shiny; no fossils left	building stone; statues; work surfaces
mudstone (layers; soft and crumbles easily)	slate – very hard, shiny; splits in a single direction to give flat sheets	roofing; facings for buildings

3. Explain how metamorphic rocks differ from sedimentary rocks.
4. Suggest why there are no fossils (or only very distorted ones) in metamorphic rocks formed from sedimentary rocks.

Metamorphic rocks in detail

Formation of metamorphic rocks varies a great deal depending on the temperature and pressure applied. Each set of conditions produces different rocks. The most intense metamorphism is called high-grade metamorphism. It produces gneiss (pronounced 'nice'), which has alternating bands of light and dark minerals. This type of metamorphism is often associated with the collision of tectonic plates and the formation of new mountains.

Heat and high pressure can destroy information contained in rocks. Limestone that is full of marine fossils may metamorphose into marble that is fossil-free. The heat (and pressure) destroys the fossils and hence the clues to the origin of the rock.

5. Why can many different metamorphic rocks be formed from the same sedimentary rock?
6. Metamorphic rocks do not usually provide geologists with much evidence about the past. Explain why not.
7. Ceramics are materials that are used to make products such as wash-hand basins and crockery.
    a) Suggest similarities and differences between ceramics and metamorphic rocks.
    b) Suggest whether ceramics are more similar to metamorphic or to other types of rock.

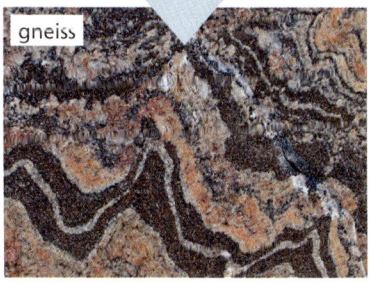

FIGURE 1.7.4b: The effect on shale (a sedimentary rock) of exposure to more and more heat and pressure. The shale metamorphoses, via slate, phyllite and schist into gneiss.

Know this vocabulary

metamorphic
metamorphose

SEARCH: metamorphic rocks

Earth

Understanding the rock cycle

We are learning how to:
- Describe the rock cycle.
- Explain how rocks can change from one type to another.

The three main types of rock on Earth are all related and the amount of each type changes constantly. Which processes link the different rocks?

The rock cycle

The Earth's rocks are continually changing because of processes such as weathering, erosion and large Earth movements. The rocks are slowly recycled into other types over millions of years – this is known as the **rock cycle**.

The movement of tectonic plates and the Earth's inner heat drive the rock cycle. Look at Figure 1.7.5a, which illustrates the processes in the rock cycle. Mountains and hills form when buried rocks are moved to the surface. This is called **uplift**. Rocks at the surface are weathered and pieces break off. Erosion occurs when rock particles are worn away and moved elsewhere.

1. How are rocks changed from one type to another?
2. What is the difference between weathering and erosion?

Did you know...?

A Scottish scientist called James Hutton found evidence that the Earth had experienced extremely high pressures – enough to uplift and tilt rocks – and temperatures high enough to melt rocks and drive the rock cycle we understand today. He is recognised as the founder of modern geology.

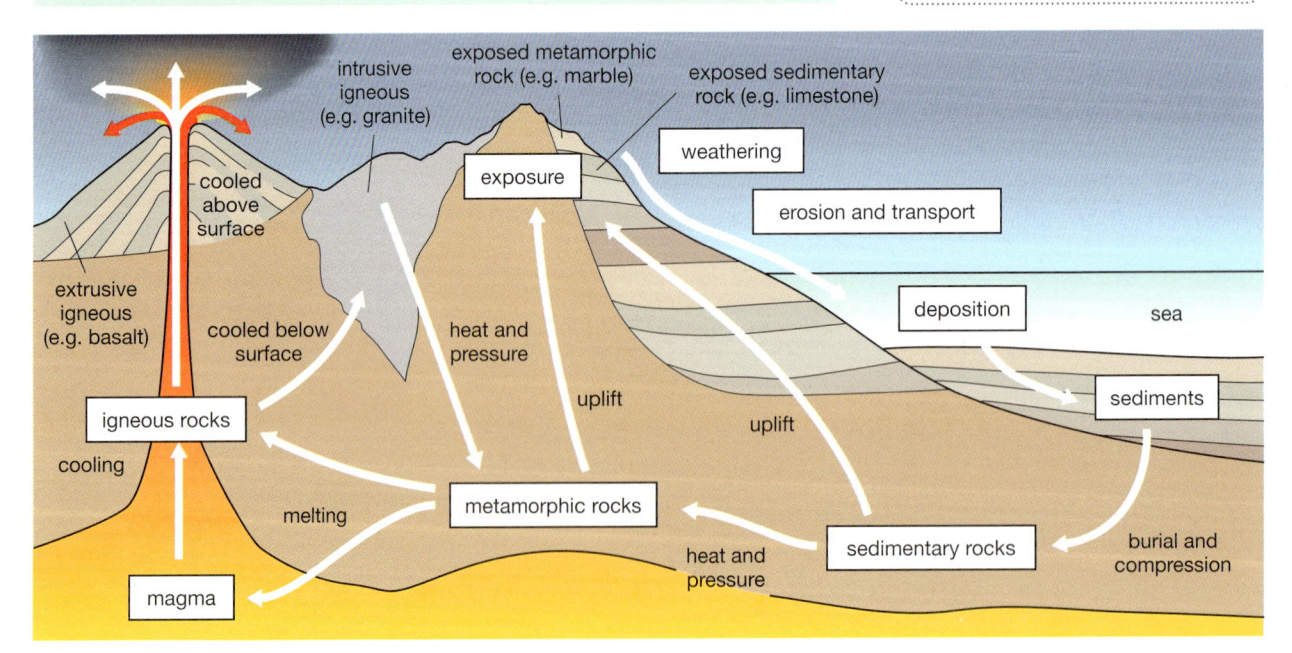

FIGURE 1.7.5a: Uplift causes the continual movement and cycling of rocks.

160　AQA KS3 Science Student Book Part 1: Earth – Earth structure *and* Universe

Folding rocks

7.5

Earth movements can squeeze layers of rock into massive folds, forming mountains – for example the Alps, the Rockies and the Zagros mountains. The Alps are so old that the top halves of the folds have been worn away by the weather.

Folding can be seen on a small scale in coastal cliffs (Figure 1.7.5b). When strata (layers of rock) are pushed up into a dome shape it is called an **upfold** or **anticline**. When strata are forced down into a bowl shape it is called a **syncline**.

Sediments and lava flows are usually deposited in horizontal layers due to the effect of gravity. Earth movements cause these layers to bend, tilt or fracture into pieces.

FIGURE 1.7.5b: Folded strata, clearly visible at Stair Hole cliffs in Dorset.

3. Explain the part these processes play in the rock cycle:
    a) erosion;
    b) deposition;
    c) heat and pressure.
4. Explain how anticlines and synclines form.

Rocks on the move

The movement of the Earth's crust causes rocks deep underground to be brought up to the Earth's surface, in the process of uplift. Uplift is occurring continually in some areas of the world today, such as Taiwan.

Faulting occurs when rocks break because of the forces acting on them. Stress builds up over years until the rocks move. Rocks can move from a few centimetres up to a few metres. When this happens, huge amounts of energy are released in earthquakes.

FIGURE 1.7.5c: Taiwan is rising by over 1 cm every year. During a big earthquake in 1999, uplift of up to 9.5 metres in some places caused significant damage to many large buildings and structures.

5. Explain how rocks from deep underground can quickly reach the surface of the Earth.
6. Which of the processes in the rock cycle happen quickly and which happen slowly?

Know this vocabulary

rock cycle
uplift
upfold
anticline
syncline

SEARCH: the rock cycle

Earth

Describing stars and galaxies

We are learning how to:
- Describe the characteristics of a star.
- Relate our Sun to other stars.
- Explain the concept of galaxies.

Our Sun is the star that maintains the conditions that allow life to exist on Earth. It sits in a galaxy called the Milky Way. The Milky Way is one of over 170 billion galaxies in the Universe. Each galaxy contains billions of stars.

Characteristics of a star

A **star** forms when a huge cloud of matter (usually hydrogen) is pulled together by its own gravity. Eventually the temperature and pressure become so high that the hydrogen atoms join to make helium. This process, known as **nuclear fusion**, releases the huge amount of energy that makes a star shine so brightly.

Our Sun is quite a small star. However, it has a diameter 109 times that of the Earth and it contains 99.9 per cent of the matter in the solar system.

FIGURE 1.7.6a: The Sun – huge amounts of energy are released by nuclear fusion, including light. It takes light around eight minutes to travel from the Sun to the Earth.

1. What are the main chemical elements in the Sun?
2. Describe where the Sun gets its energy from.
3. Thinking about energy, suggest the main difference between a star and a planet.

Different types of stars

The size and age of a star determine its characteristics. Figure 1.7.6b shows two of the brightest stars in the night sky – Rigel and Betelgeuse. At about 57 times the diameter of the Sun, Rigel is a blue-white supergiant that shines with an intensity more than 50 000 times larger than the Sun. The extremely high rate at which it is fusing hydrogen accounts for its brilliance. Betelgeuse is a bigger, older star known as a **red giant**. It has run out of hydrogen so is now fusing helium atoms as its source of energy. Its average diameter is about 950 times that of the Sun.

When a star much larger than our Sun approaches the end of its life, its inner core can collapse to form a **neutron star**. A neutron star has a mass similar to that of the Sun, concentrated into a diameter of about 10 km.

FIGURE 1.7.6b: Rigel and Betelgeuse are stars in the constellation of Orion.

7.6

4. Suggest what the surface gravity of a neutron star would be like.
5. Explain why stars do not have an infinite life span.

Stars and galaxies

With the naked eye it is only possible to see a tiny fraction of the stars. Even our closest stars, such as Proxima Centauri and Sirius A, are approximately 100 000 000 000 000 km away from the Earth. In scientific notation this is written as 10^{14} km.

Our solar system and our closest stars are part of a **galaxy** called the Milky Way, which is similar in shape to the Andromeda galaxy shown in Figure 1.7.6c. Galaxies are so large it takes many years for light to travel across them, and billions of years for it to travel between galaxies.

Evidence about space is collected through telescopes and through analysis of the light that reaches the Earth. Scientists have shown that since the Universe was created in the Big Bang, it has been continually expanding. By analysing the light that arrives on Earth from distant galaxies, scientists are able to measure the rate of expansion.

> **Did you know…?**
>
> We are now discovering planets that orbit stars other than our Sun. These are called **exoplanets**.

FIGURE 1.7.6c: Andromeda galaxy – the bright haze consists of many distant stars.

6. Explain why scientific notation is sometimes used for writing numbers.
7. Explain the differences between these terms, and list their order of size: star, Universe, planet, galaxy.
8. Our ideas about the universe have sometimes changed – why aren't new explanations always immediately accepted?

Know this vocabulary

star
nuclear fusion
red giant
neutron star
galaxy
exoplanet

SEARCH: stars and galaxies

Earth

Explaining the effects of the Earth's motion

We are learning how to:
- Describe variation in length of day, apparent position of the Sun and seasonal variations.
- Compare these with changes in the opposite hemisphere.
- Explain these changes with reference to the motion of the Earth.

The Earth's rotation defines day length. The time taken for the Earth to orbit the Sun defines the length of a year. Because the orbit takes 365.25 days, we have an 'extra' day every leap year. Without this the seasons would drift – after 730 years, midsummer would be in December.

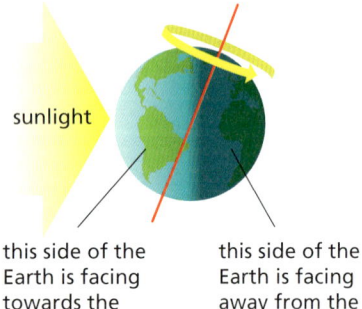

the Earth spins around its axis every 24 hours

sunlight

this side of the Earth is facing towards the Sun – it is day here

this side of the Earth is facing away from the Sun – it is night here

Day and night

Look at Figure 1.7.7a. On the side of the Earth that is facing the Sun it is day; and on the opposite side it is night.

Figure 1.7.7b shows how the length of daytime varies throughout the year at two locations.

FIGURE 1.7.7a: Day and night are caused by the Earth spinning on its axis.

1. Look at Figure 1.7.7b. Compare the length of daytime in the Arctic Circle and in northern France on:
    a) 21 December;
    b) 21 June.
2. What is special about 21 March and 21 September?

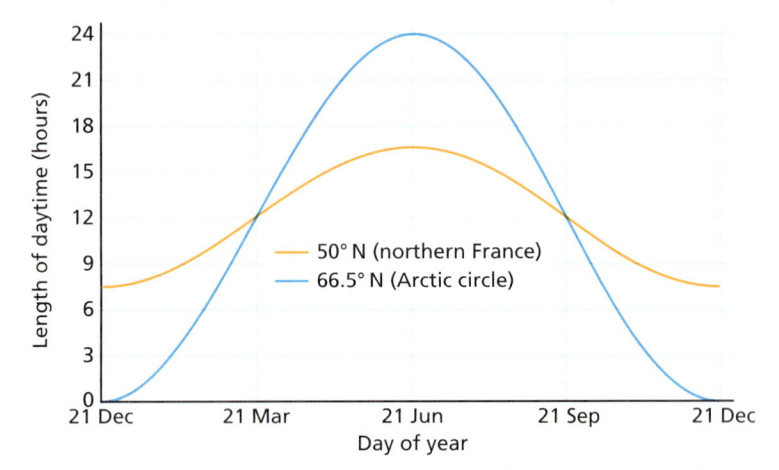

50° N (northern France)
66.5° N (Arctic circle)

FIGURE 1.7.7b: Variation in length of daytime.

Tilt of the Earth's axis

The Earth's **axis of rotation** is tilted at 23°. Figure 1.7.7c shows this tilt and how the Earth orbits the Sun once a year. When the northern hemisphere is tilted away from the Sun, the daytimes are shorter. The Sun is low in the sky, even at midday and the amount of heat from the Sun is reduced – it is the **season** of winter. Six months later the Earth is on the other side of the Sun, which means that the northern hemisphere is now angled towards the Sun – it is summer.

3. How would the graph look different for locations in the southern hemisphere?

164 AQA KS3 Science Student Book Part 1: Earth – Earth structure *and* Universe

4. Explain why the daytimes are longer than the night-times during summer.
5. At what times of the year are daytimes and night-times of equal length? Explain why this happens.
6. Explain the changes in seasons and day length a country on the equator experiences.

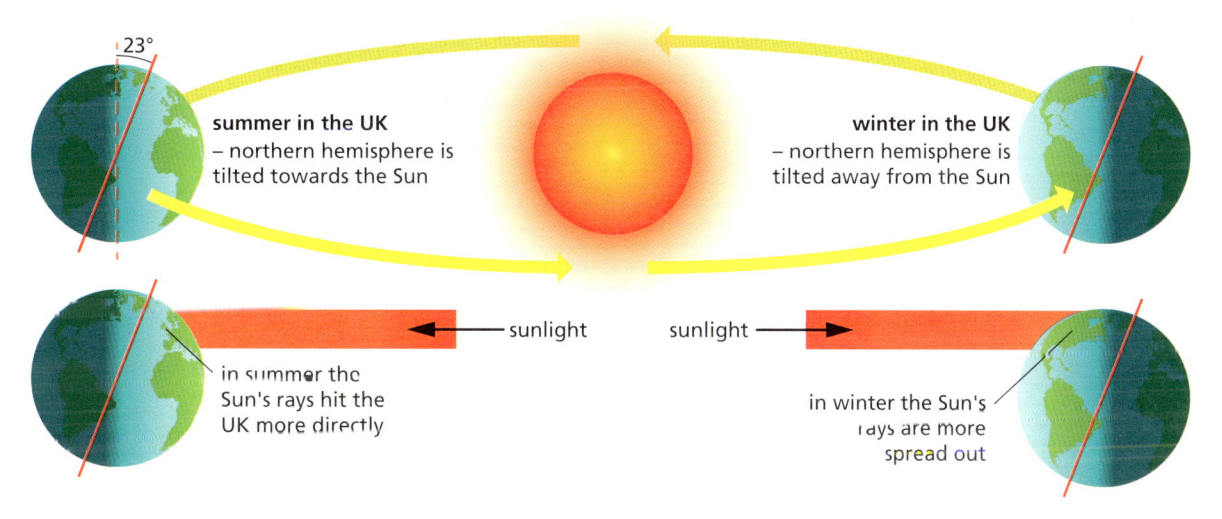

7.7

FIGURE 1.7.7c: The seasons are caused by the Earth's tilt.

Implications of the Earth's tilt

If the tilt angle of the Earth's axis were zero, we would not experience seasons as we currently do. There would be only small annual variations because of the Earth's slightly elliptical orbit, meaning that the Earth is closer to the Sun at some times than at others.

If the Earth's axis were tilted more than its current 23°, it would make the seasonal variations more extreme. At places where the Sun is directly overhead, the amount of energy reaching that place is at a maximum. Where the Sun's rays meet a place at an angle, the available energy is spread out over a larger area.

7. At midsummer on Earth, the Sun never sets at the poles – explain why. Draw diagrams to help.
8. Explain why plants grow less well in the winter than they do in the summer.
9. Mars is tilted more on its axis and has a much more elliptical orbit than the Earth. It spins at about the same rate as the Earth, but takes twice as long to complete one orbit. From this information, suggest how Mars' days, seasons and years might differ from those on Earth.

Did you know...?

The planet Uranus has a unique feature in the solar system – its axis is tilted at 82°, meaning it rotates 'on its side' as it orbits the Sun. Its orbit takes 84 Earth years.

Know this vocabulary

axis of rotation
season

SEARCH: causes of day and night and seasons

Earth

Exploring our neighbours in the Universe

We are learning how to:

- Recall that the light year is used to measure astronomical distances.
- Explain the limitation of units such as km in describing astronomical distances.
- Explain what causes the appearance of the Moon to change.

The distance from the Earth to the Sun is about 150 million km. This distance is tiny compared to distances to other stars. Dealing with such vast distances is difficult, so special units are needed.

FIGURE 1.7.8a: The light from some stars has taken many millions of years to reach us.

Light years

When measuring distances across the Universe we often use the **light year** (or ly for short). It gives more manageable numbers than using kilometres. The unit is defined by how far light will travel in a year. When travelling through a vacuum, light has a speed of just under 300 000 km/s. This means that in 1 year, light will travel 9 460 000 000 000 km through space – this is how many km there are in 1 ly.

1. What does the abbreviation 'km/s' mean?
2. What unit is often used to measure distances across the Universe?
3. Explain why distances across the Universe are not normally measured in kilometres.

Distances in the Universe

Even when measuring in light years it still does not stop the distances involved being almost mind-numbing, for instance, look at Table 1.7.8. Even when Pluto is at its closest to the Earth, it is about 300 times further away from us than the Sun. However, the distance to Pluto from the Earth is minuscule compared to distances to other stars. Light from the far reaches of the Universe has taken 15 billion years to reach the Earth. This means that we are seeing those places as they were 15 billion years ago. It also means that manned missions would have to cope with journeys lasting long periods of time, such as years or even decades.

Did you know…?

As well as using light years to measure distance, astronomers use other units – the astronomical unit (AU), where 1 AU = the mean distance between the Earth and the Sun, and the parsec (pc), where 1 pc = 3.26 ly.

7.8

Examples in the Universe	Distance (ly)
Distance across the Milky Way galaxy	100 000
Earth to Sirius (one of our nearest stars)	8.6
Earth to the most distant point of the Universe	15 000 000 000
Earth to Pluto (at their closest)	0.000 44
Earth to the Sun	0.000 016
Earth to the Andromeda galaxy	2 500 000

TABLE 1.7.8: Some approximate distances in the Universe.

4. If you were to look at Sirius today, in which year would the light you see from it have set off?
5. When you look at two stars that are different distances away, you are not seeing them at the same point in time – explain why.
6. Why do these distances have implications for exploration of the other parts of the Universe?

How planets and moons look

Stars make their own light but planets and moons reflect the light. This means that only one side of them is lit at any one time and this changes their appearance when viewed from elsewhere. The Moon orbits the Earth but reflects light from the Sun so one half of the Moon is lit. We often don't see it as a complete sphere. If we can see all the lit side, we call that a full Moon. As the Moon orbits the Earth, we see less of the lit side until it becomes a crescent. The next stage is for the dark side to be facing us (new Moon) and it then gradually shows us more of the lit side again. These shapes are called phases of the Moon.

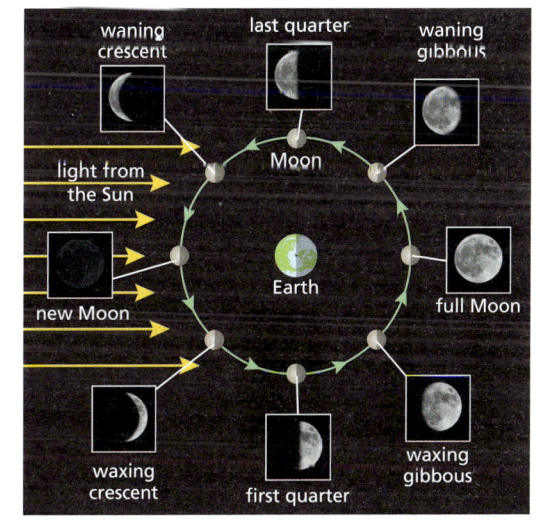

FIGURE 1.7.8b: The phases of the Moon.

7. If we can see half of the lit side, what shape will the Moon appear to be?
8. Why is one of the crescent Moons labelled 'waning' and the other 'waxing'?
9. Explain whether, if you lived on the Moon, you would see phases of the Earth.

The same thing applies to planets as well. They are lit on one side by the Sun. Figure 1.7.8c shows how Venus appears from the Earth.

10. Draw and label diagrams to suggest why:
    a) we also see phases of Mercury;
    b) we don't see a full set of phases for planets in the solar system further out than Earth.

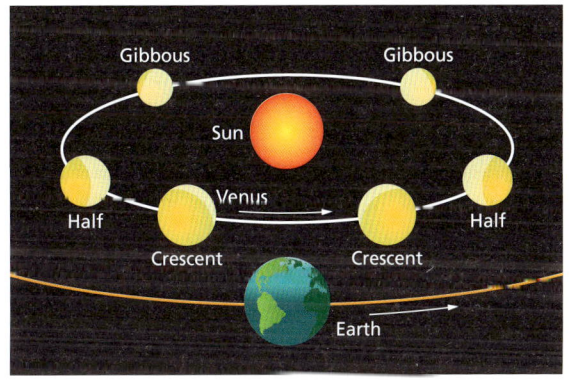

FIGURE 1.7.8c: The phases of Venus.

Know this vocabulary

light year (ly)

SEARCH: distances in the Universe

Earth

Using models in science

We are learning how to:
- Explore how we can use models to explain ideas in science.
- Construct an explanation using ideas and evidence.
- Decide if a model is good enough to be useful.

People make models for a variety of different reasons. If a town planner has some ideas about a new shopping centre they are proposing, they may build a model to show people what it would look like. It's a good way of getting people to see what might be good about it and also what problems it might cause. Scientists sometimes use models to explain ideas but more often to help them develop their ideas. Instead of drawing a conclusion and then making a model to illustrate it, they might do it the other way around – using a model helps them to draw a better conclusion.

FIGURE 1.7.9a: Students examining a model of the Earth and the Sun.

Modelling day length

A team of students is trying to come up with a **model**. Their teacher has asked them to devise a way of explaining why in the UK we get longer days in the summer and shorter days in the winter, but in the southern hemisphere it's the other way around. They've got a torch to represent the Sun and a model Earth on a knitting needle to represent the Earth. They're trying to work out how to position them.

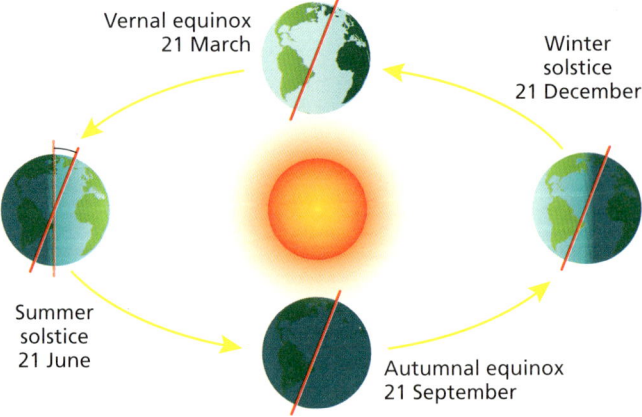

Vernal equinox 21 March
Winter solstice 21 December
Summer solstice 21 June
Autumnal equinox 21 September

FIGURE 1.7.9b: Why does day length change over the year?

They've worked out that they have to tilt the needle. They angle the top of the Earth (the northern hemisphere) away from the torch and, as the Earth spins, it doesn't get light for as long. This is winter. They're not quite sure how to model summer though. Then they find a diagram in a book.

1. What does 'northern hemisphere' mean?
2. How does the students' model show that when it's winter in the northern hemisphere it's summer in the southern hemisphere?
3. How should the team show what will be happening six months later?

7.9

Modelling types of rock

A science teacher has found a great way of showing the difference between sedimentary, metamorphic and igneous rocks – she is going to get the class to use chocolate! First they're going to model sedimentary rocks. They grate some chocolate, wrap it in aluminium foil and squash it, really hard. Imagine what that will look like when they open it out and examine it.

Next they model metamorphic rock. They use some of the chocolate shavings left over along with some lumps and put them in a paper bun case. They float this on hot water and when it starts to melt take it out. They have to be careful examining this one because it's hot.

The last one they do is igneous rock. The teacher gathers up the various pieces of chocolate from the previous experiments and puts them in a foil dish, which she floats on hot water. This time it's kept there for several minutes until it's totally melted.

FIGURE 1.7.9c: Grated chocolate can be used to model different types of rock.

4. In what way will the sedimentary model look like a sedimentary rock?
5. How does the metamorphic model represent what happens to rocks that go through that process?
6. Explain how good the model for making igneous rocks is.

Scientists as modellers

Both of these models, the one showing seasonal change and the one showing rock processes, can be used to help us to understand the processes better. Sometimes we use a model because it's a convenient size, such as the one about seasons, and sometimes because it's safer and more convenient. Jake's teacher had to be careful with the molten chocolate but it was safer than lava.

However, a model represents only certain features, and we have to decide how well it represents them.

7. With the Sun–Earth model:
   a) Which features of the real Sun–Earth system did it represent well?
   b) Which features did it not represent well?
8. Evaluate the models of each type of rock-forming process using chocolate, clearly identifying the good features and the aspects less well represented.

Did you know…?

Using models is important in many areas of work. Climate scientists use models of the atmosphere. The structure of DNA was explained by building a model to show the double helix. Town planners use models to predict traffic flow, but such a model is difficult to get right because it depends on human behaviour.

Know this vocabulary

model

Earth

Checking your progress

To make good progress in understanding science you need to focus on these ideas and skills.

- ☐ Name the layers that make up the Earth and recall that the Earth's surface is made of plates that move about.

- ☐ Describe the characteristics of each layer of the Earth and recall that tectonic plates move very slowly.

- ☐ Explain that earthquakes, volcanic eruptions and the formation of mountains can happen where tectonic plates meet; explain how volcanic activity changes the surface of the Earth.

- ☐ Describe how igneous, sedimentary and metamorphic rocks are formed; give examples and describe how they can change from one type to another.

- ☐ Describe the features and properties of different types of rocks, including crystals in igneous rocks, recrystallisation in metamorphic rocks and layers (burying fossils) in sedimentary rocks.

- ☐ Explain the processes involved in the rock cycle and link these to how the rocks are formed.

- ☐ Describe what is meant by weathering and erosion.

- ☐ Identify causes of weathering and erosion.

- ☐ Explain how weathering and erosion affect rocks.

7.10

- [ ] Describe the relative motion of the Earth, Moon and Sun.
- [ ] Explain how the motion of the Earth relative to the Sun causes day length and year length.
- [ ] Explain how the relative motion of the Earth, Moon and Sun affects how we see objects from the Earth.

- [ ] Explain how the Earth is tilted upon its axis.
- [ ] Explain how the tilt of the Earth on its axis causes seasonal changes.
- [ ] Explain the effects of the tilt on a planet's axis being greater or less.

- [ ] Recall the time taken for light to reach Earth from the Sun and from the next nearest star.
- [ ] Explain the choice of units used for measuring distances in space.
- [ ] Explain how observations of stars are affected by the scale of the Universe.

- [ ] Describe what a galaxy is.
- [ ] Explain what has been learned from the observation of galaxies.
- [ ] Explain the importance of the discovery of exoplanets.

Earth

Questions

KNOW. Questions 1–5

See how well you have understood the ideas in this chapter.

1. Which is the correct order of the layers of the Earth, starting from the centre? [1]
   a) Mantle, inner core, outer core, crust.
   b) Outer core, inner core, crust, mantle.
   c) Inner core, outer core, crust, mantle.
   d) Inner core, outer core, mantle, crust.

2. Which statement describes the mantle correctly? [1]
   a) A relatively thin, rocky layer.
   b) A very thick layer, some of which can flow.
   c) Made of liquid nickel and iron.
   d) Made of solid nickel and iron.

3. In the summer, in the UK, the days are longer and the average temperatures higher. Which of these statements explains why? [1]
   a) In summer the Earth is closer to the Sun.
   b) In summer the northern hemisphere is tilted towards the Sun.
   c) The Earth's orbit is not exactly circular.
   d) A long hot summer is needed to help plants grow.

4. It takes around eight minutes for light to reach the Earth from the Sun. Which of these statements is *not* true? [1]
   a) The Sun appears in the sky to us where it was eight minutes previously.
   b) Light from the stars takes much longer to reach us, often taking many years.
   c) It takes eight minutes for light from all the stars to reach the Earth.
   d) Light from the Sun takes different amounts of time to reach the different planets in the solar system.

5. The Moon is sometimes visible in the sky and sometimes not. Which of these statements is true? [1]
   a) The Moon orbits the Sun and is sometimes therefore much further away from us.
   b) During the day the Moon is on the other side of the Earth; its orbit brings it in view at night time.
   c) Sometimes the Moon doesn't generate enough light to be seen.
   d) Seeing the Moon depends on where it is in its orbit around the Earth and on the rotation of the Earth.

APPLY. Questions 6–8

7.11

See how well you can apply the ideas in this chapter to new situations.

6. Why does magma come out of volcanoes? [1]
   a) It is less dense than the crust.
   b) Uplift forces it out.
   c) An explosion forces it out.
   d) It is more dense than the crust.

7. Scientists find a new type of rock that has small crystals. What does this tell them? [2]

8. In its orbit around the Sun, the Earth is tilted on its axis by about 23°. Explain what we would notice happen about seasons and climate if it were tilted at a greater angle. [2]

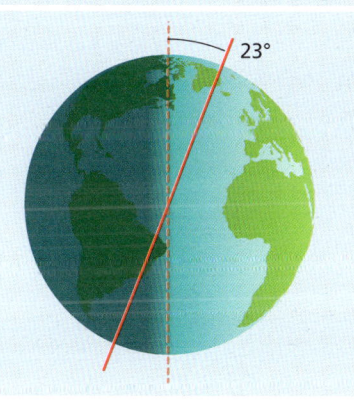

FIGURE 1.7.11a

EXTEND. Questions 9–10

See how well you can understand and explain new ideas and evidence.

9. Why does the water content of new sedimentary rock change as the rock forms? [2]

10. Table 1.7.11 shows the planets in the solar system, their masses and the number of moons they have. Suggest a pattern between the mass and the number of moons. [2]

Planet	Mass (units are 10^{24} kg, which is a shorthand way of writing 1,000,000,000,000,000,000,000,000 kg)	Number of moons
Mercury	0.33	0
Venus	4.87	0
Earth	5.97	1
Mars	0.64	2
Jupiter	1898	67
Saturn	569	62
Uranus	86.8	27
Neptune	102	14

TABLE 1.7.11

Organisms
Movement *and* Cells

Ideas you have met before

Movement

Humans and some other animals have a skeleton to support and protect them.

Animals with a backbone are called vertebrates.

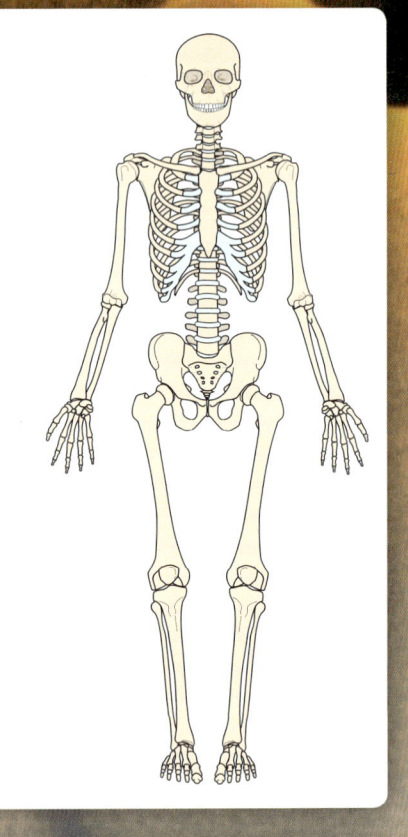

Body systems

We can think of a human body as being made up of different systems.

Each system has a specific purpose in the body.

We have a circulatory system that pumps blood around, a skeletal system that supports us and a digestive system that provides energy from the food we eat.

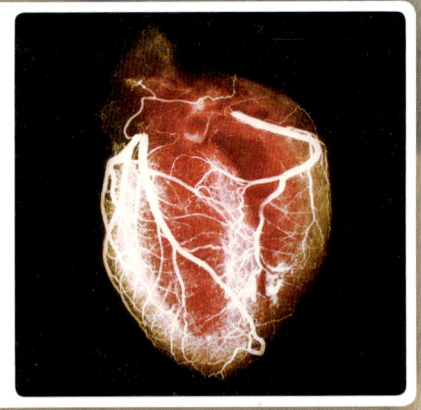

In this chapter you will find out   **8**.0

The skeleton
- The skeleton allows movement at the joints.
- The skeleton also protects some organs.
- Most blood cells are made inside bones.

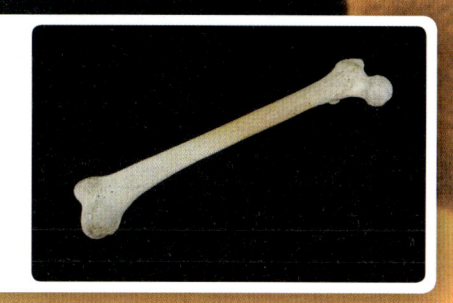

Muscles
- Muscles contract to move bones at the joints.
- Muscles can only contract and relax – they cannot push.
- Many muscles interact and work in pairs to bring about opposite movements.

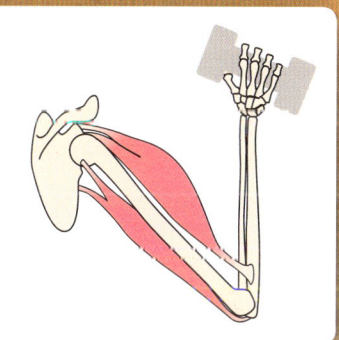

Cells
- Cells are the building blocks of life. They contain structures called organelles, which all have specific jobs.
- Microscopes can be used to observe cells and other structures.
- Some organisms, such as bacteria and protozoa, consist of only a single cell. They can, nevertheless, carry out all necessary life processes.

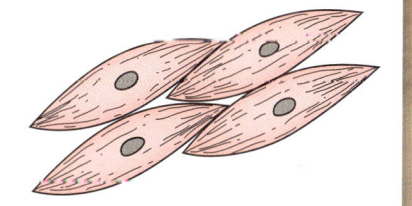

How cells work for an organism
- A human body has a highly organised set of organ systems, organs, tissues and cells.
- Many cells, such as muscle cells and nerve cells, are specialised, enabling them to carry out a specific task more effectively.
- Body systems can be affected by certain drugs and by damage to other organs.

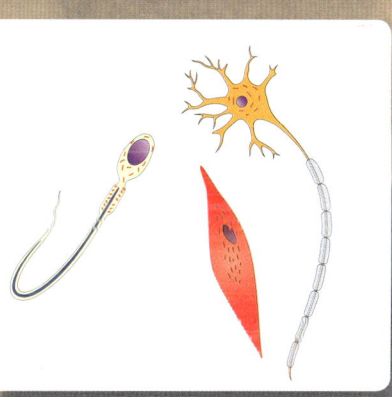

175

Organisms

Exploring the human skeleton

We are learning how to:
- Identify bones of the human skeleton.
- Describe the roles of the skeleton.
- Explain why we have different shapes and sizes of bones.

There are 206 bones in the human skeleton. Each one contains calcium to make it strong. The smallest bone is found in your ear and is approximately 3 mm long. The largest is your thigh bone. Why do bones vary so much?

> **Did you know…?**
>
> Bones of flying birds are hollow to make the skeleton more lightweight. To increase the strength of the skeleton, more bones are fused together than in humans.

The human skeleton

Bones make up the human **skeleton**.

Look at Figure 1.8.1a and answer these questions.

1. State the scientific name for the:
   a) skull;
   b) collar bone;
   c) shoulder blade;
   d) funny bone.

2. Suggest why you cannot count 206 bones on the diagram of the skeleton in Figure 1.8.1a.

3. Explain why the name 'vertebrates' is suitable for describing animals that have a backbone. **Hint:** look at the bones of the backbone.

Roles of the skeleton

The human skeleton has four main roles:
- it supports the body;
- it protects the organs;
- it allows movement;
- it produces blood cells.

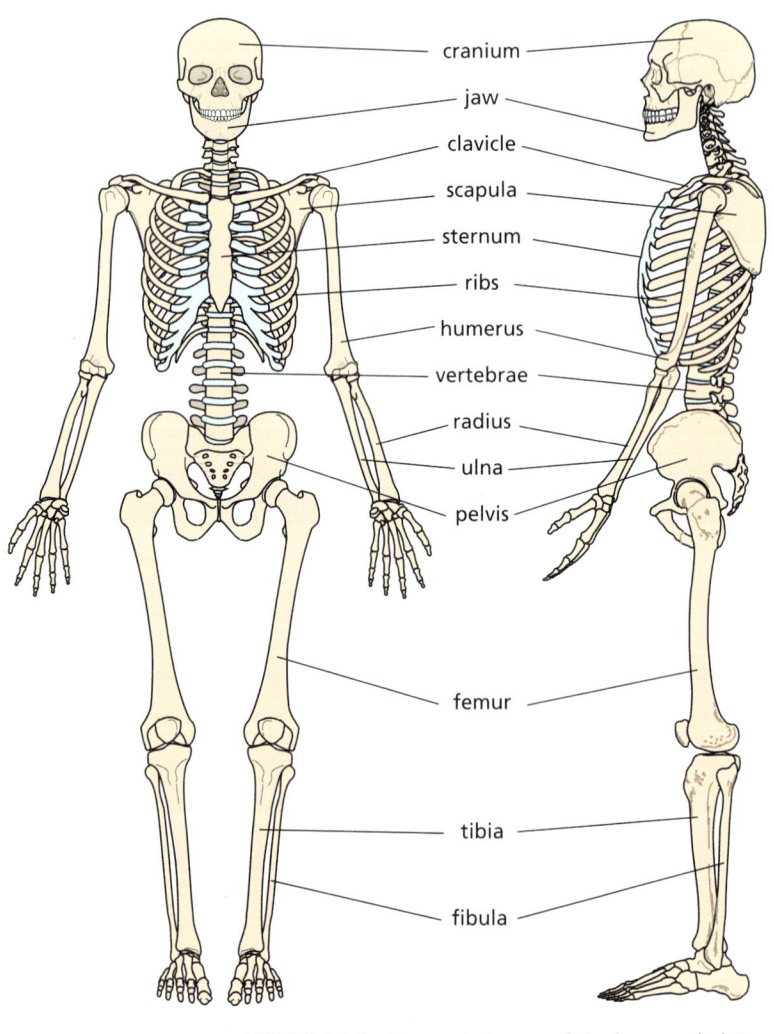

FIGURE 1.8.1a: The main bones of the human skeleton.

Labels: cranium, jaw, clavicle, scapula, sternum, ribs, humerus, vertebrae, radius, ulna, pelvis, femur, tibia, fibula

176   AQA KS3 Science Student Book Part 1: Organisms – Movement *and* Cells

8.1

Without a skeleton you would not be able to sit, stand or hold yourself up.

The ribs are curved bones, forming a cavity inside the ribcage. The lungs are positioned inside the ribcage.

The many **joints** in your skeleton allow you to move. For example, the joint at the knee allows your leg to bend.

New blood cells are made in the **bone marrow** (Figure 1.8.1b).

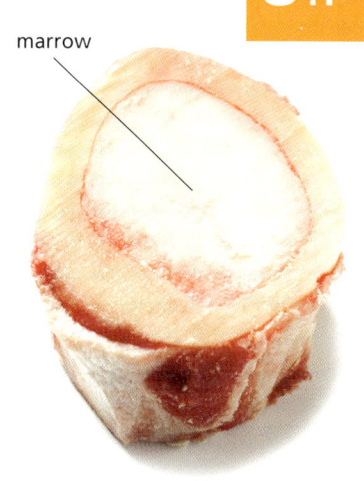

> 4. Describe the four main roles of the skeleton.
> 5. Explain which organ each part of the skeleton protects:
>   a) ribs;
>   b) cranium.
> 6. Describe three parts of the skeleton where joints are important.

FIGURE 1.8.1b: Bone marrow inside large bones makes blood cells. The red blood cells carry oxygen around the body.

Comparing bones

Bones must be strong to support you. Most of a bone (approximately 70 per cent) is made up of hard minerals, such as **calcium**. The outside of a bone is smooth and hard to provide support. Inside this hard outer layer lies spongy, porous material and inside this layer, in some bones, is bone marrow. This makes your bones lighter than if they were completely solid.

Some bones are long and narrow, such as those in your legs. Some bones are shorter, such as those in your feet. Other bones are flat and wide, such as the scapula (shoulder blade). Each bone is adapted to suit its function. For example, the foot contains many small bones to allow flexibility.

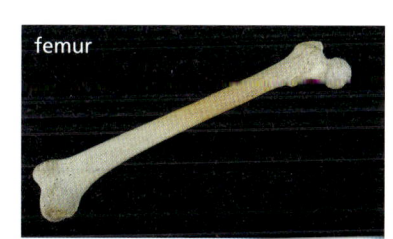

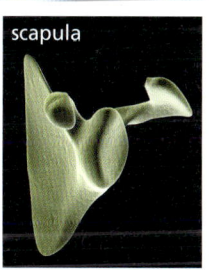

FIGURE 1.8.1c: The femur is a long bone, whereas the scapula is a flat bone.

> 7. For each of the examples below, describe how the bone shape or structure is well adapted for its function in the body:
>   a) femur (thigh bone);
>   b) bones of the hand;
>   c) ribs.
> 8. Vertebrae are described as small and irregular bones.
>   a) Explain what is meant by an 'irregular' bone.
>   b) Suggest why all vertebrae are small and the same size.

Know this vocabulary

skeleton
joints
bone marrow
calcium

SEARCH: the human skeleton   177

Organisms

Understanding the role of joints and muscles

We are learning how to:
- Describe the roles of tendons, ligaments, joints and muscles.
- Identify muscles used in different movements.
- Compare different joints in the human skeleton.

Bones meet at joints. Some joints, such as those in your cranium, do not allow much movement. However, many joints allow a wide range of movement. Try moving your arm at your elbow, then try at your shoulder. Different joints allow you to move in different ways. Muscles cause these movements at joints by pulling the bones.

Tendons and ligaments

The bones of a skeleton are held together by **ligaments**. Bones are connected to **muscles** by **tendons**.

Both ligaments and tendons are made of fibres called collagen. However, the fibres are arranged differently in each. In tendons, they are arranged so that the tendon can move easily as muscles contract. In ligaments, fibres are arranged more tightly to hold bones together securely.

1. Describe the roles of tendons and ligaments.
2. Sportspeople often damage ligaments. Suggest how this can happen.

The main muscles of the body

There are three types of muscle – cardiac muscle in the heart, smooth muscle in the organs, and skeletal muscle attached to the skeleton.

Skeletal muscles (Figure 1.8.2b) allow you to move. They are attached to bones by tendons. As the muscles contract, they pull on tendons, causing the bones around a joint to move. You have over 600 skeletal muscles, which are all involved in moving parts of your body.

The skeletal bones, ligaments, skeletal muscles and tendons are collectively called the **muscular skeletal system**.

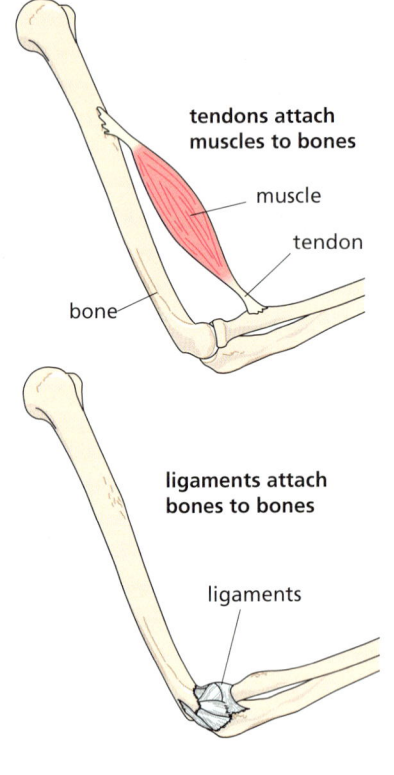

FIGURE 1.8.2a: Tendons join bone to muscle; ligaments join bone to bone.

Did you know…?

The heart is made of muscle. But this is different from the muscles attached to your skeleton. Heart muscle (cardiac muscle) contracts approximately 70 times every minute for your entire life and it does not tire.

8.2

3. State the three types of muscle and where each is found.
4. Name three muscles of the arm.
5. Describe the main muscles that enable you to make these movements:
   a) reaching your arms above your head;
   b) lifting your toes off the floor;
   c) doing a sit-up.

Are all joints the same?

We have three types of moveable joint. The type of movement that they allow varies.

Ball and socket joints allow forward, backward and circular movements. The hip joint is a ball and socket.

A hinge joint allows movement like the opening and closing of a door. This type of joint is found at the elbow.

A pivot joint allows rotation around an axis. This type of joint is found at the top of the neck.

There are also fixed joints, such as those in the skull, that do not allow movement.

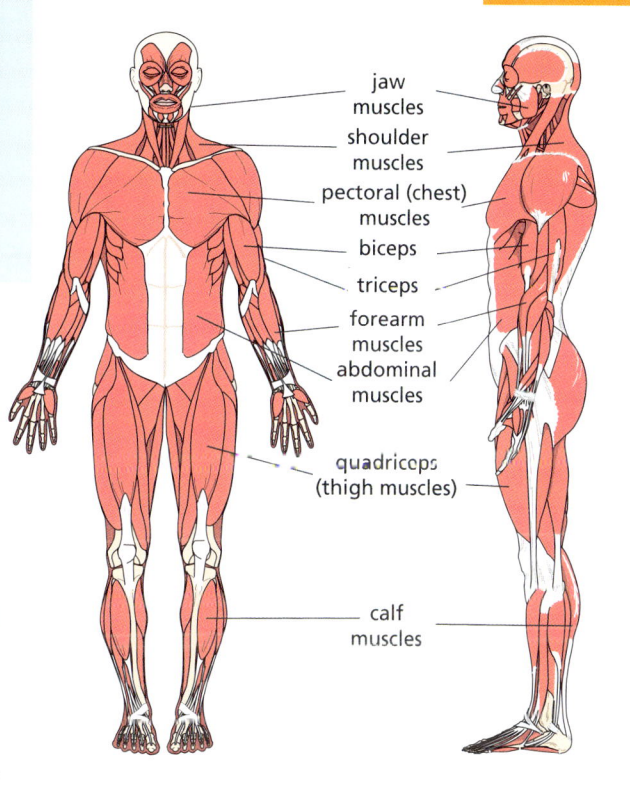

FIGURE 1.8.2b: The main skeletal muscles.

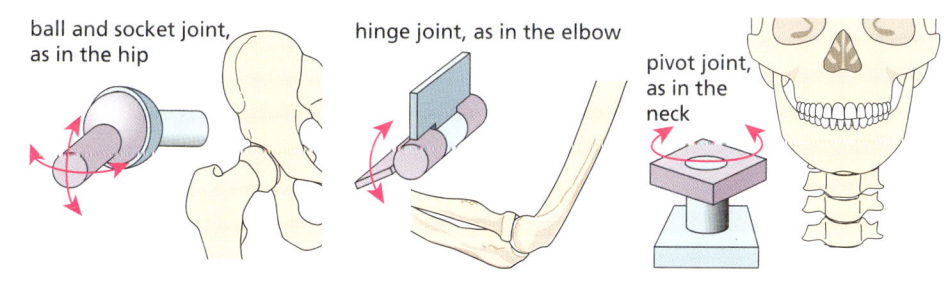

(Labels: jaw muscles, shoulder muscles, pectoral (chest) muscles, biceps, triceps, forearm muscles, abdominal muscles, quadriceps (thigh muscles), calf muscles)

FIGURE 1.8.2c: Types of moveable joints.

(ball and socket joint, as in the hip; hinge joint, as in the elbow; pivot joint, as in the neck)

At the ends of bones there is smooth, tough tissue called **cartilage**. Cartilage reduces friction between bones and allows them to slide over each other.

6. List the four types of joint in order, starting with the type allowing least movement.
7. Suggest which type of joint is found in the:
   a) shoulder;   b) knee.
8. Draw a table to summarise the types of joint and the movements they allow.

Know this vocabulary

ligament
muscle
tendon
muscular skeletal system
cartilage

SEARCH: joints and muscles   179

Organisms

Examining interacting muscles

We are learning how to:
- Describe antagonistic muscles and give examples.
- Explain, using scientific vocabulary, how antagonistic muscles bring about movement.
- Plan an investigation to compare muscle strengths.

The majority of the 600 muscles of the human body work as pairs. As one muscle of the pair contracts, the other muscle relaxes, and vice versa. Without muscles working together in this way we would not be able to move our joints freely.

Muscles working in pairs

When muscles contract, they pull on both a tendon and a bone. If the bone is at a joint, the bone will move. Muscles can only pull, they cannot push. If muscles just worked singly, once pulled the bone would simply stay in that position. To solve this problem, muscles work in **antagonistic muscle pairs**. In the arm, the **bicep** and **tricep** muscles work as an antagonistic pair to control movement at the elbow. To move the forearm up, the bicep contracts and the tricep relaxes. To move the forearm down, the tricep contracts and the bicep relaxes.

Other examples of antagonistic muscles include the quadricep and hamstring muscles in the thigh, which allow bending at the knee, and the shin and calf muscles, which allow movement at the ankle.

1. List some examples of antagonistic muscles.
2. Describe the changes in the bicep and tricep muscles as the forearm moves up and down.
3. Explain why some muscles need to work in pairs.

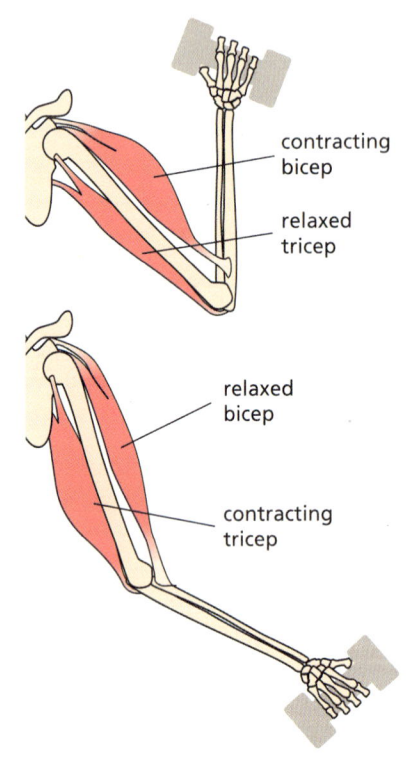

FIGURE 1.8.3a: Which way does the forearm move when the bicep contracts?

Movement in a chicken leg

The structure of a chicken leg is similar to the structure of a human leg. When we dissect a chicken leg, we can examine the tissues, including the bones, muscles and tendons. This can help us to understand how our own knee joints work by muscles pulling on the bones to cause movement.

During this dissection, tissue must be removed very carefully. If tissues such as tendons and muscles are damaged during the dissection process, it will be difficult to understand how they work together.

Did you know…?

Antagonistic muscles are at work in our eyes. Pairs of muscles in the coloured part of the eye, the iris, control how big the pupil is. This prevents the eye from being damaged by too much light entering it.

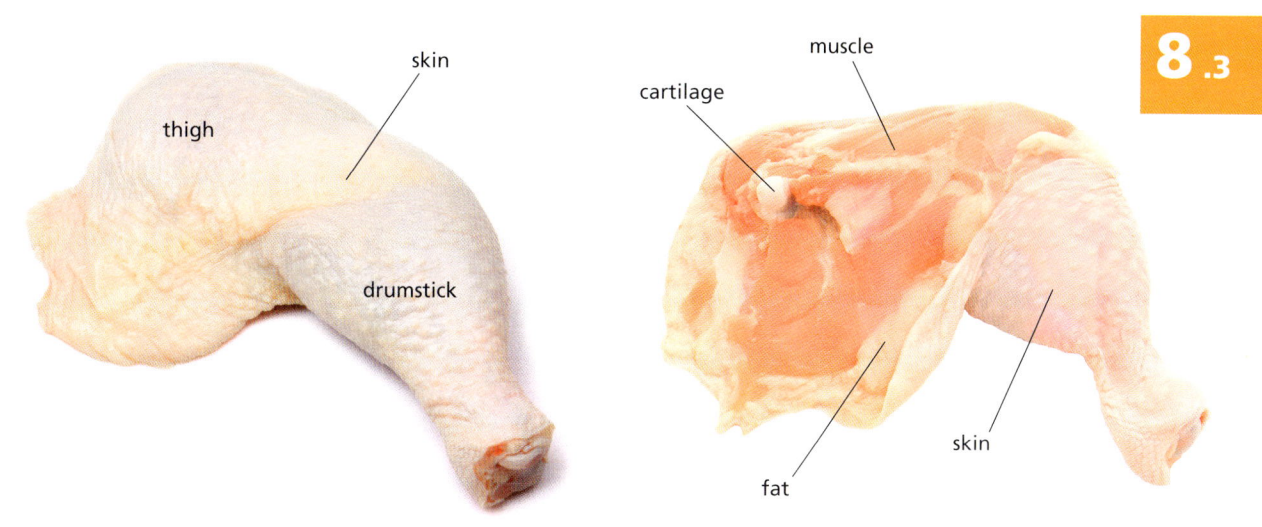

FIGURE 1.8.3b: A chicken leg.

4. Observe or carry out a dissection of a chicken leg. Describe the structure of the chicken leg and the role of each tissue. Try to include the following key words: muscle, skin, cartilage, tendon, fat, bone marrow, femur.

5. Name the type of joint found in the chicken leg and describe the movement that this type of joint allows.

6. Suggest how dissecting animal parts, such as the chicken leg, could help in human medicine.

Measuring muscle strength

By exercising you can increase the strength of muscles. Professional sportspeople consider their training very carefully to ensure that they target specific muscles.

They also test the strength of their muscles frequently to check their progress. These are scientific tests and must be carried out fairly so that measurements can be compared over time. Figure 1.8.3c shows a device to test the strength of the forearm and hand. The person squeezes the handle as hard as they can. The result is then displayed as a **force** (measured in **newtons**).

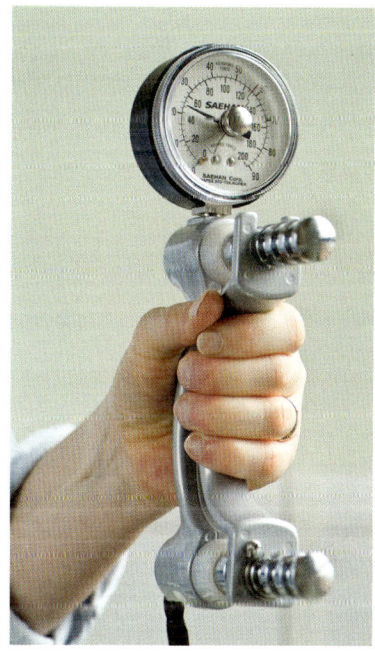

FIGURE 1.8.3c: A handgrip strength tester.

7. Describe how two rowers could compare hand and forearm strengths using a handgrip tester.

8. A basketball player wants to compare the strength of his forearm with that of a footballer. Predict who would have the most strength.

9. Suggest how you could test the strength of your quadriceps.

Know this vocabulary

antagonistic muscle pair
bicep
tricep
force
newtons

SEARCH: antagonistic muscles

Organisms

Exploring problems with the skeletal system

We are learning how to:
- Recall some medical problems with the skeletal system.
- Explain how some conditions affect the skeleton.
- Consider the benefits and risks of a technology for improving human movement.

The muscular skeletal system is made up of bones, tendons, ligaments, cartilage and muscles. Medical problems can arise with any of these components, ranging from fractures to genetic conditions that we inherit. The diagnosis and treatment of these problems have changed over time.

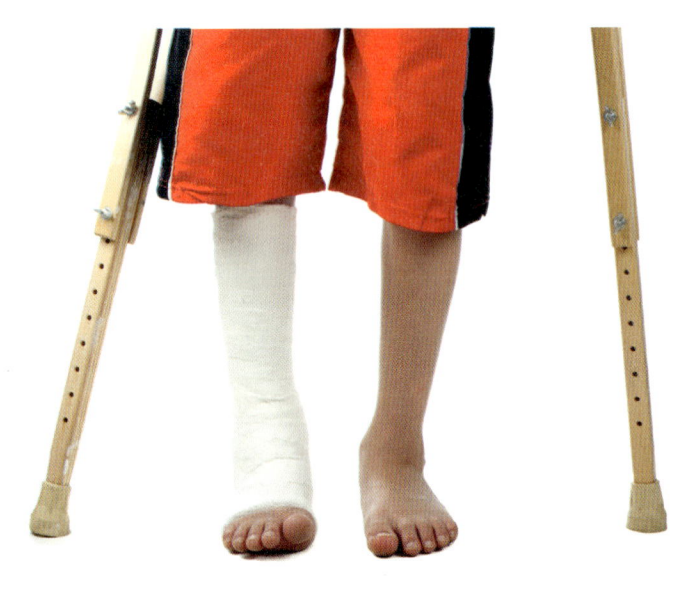

FIGURE 1.8.4a: Broken bones can heal in a cast.

Break a leg

With 206 bones in the human skeletal system, it is no surprise that bones are sometimes broken. Bones contain collagen, which allows them to bend a little. However, with a large enough impact bones can splinter, break or shatter.

Bone breaks, or **fractures**, can often be treated by covering the limb with a cast of fibreglass or plaster. This holds the bones in place while new bone knits the broken ends together. More severe fractures require metal pins through the broken bones to hold them in position while healing takes place. An open, or compound, fracture is one in which the skin is broken. This has a much higher risk of infection and usually requires surgery.

1. Suggest how a fracture may happen.
2. Describe how a fracture may be treated.
3. Explain why a compound fracture is often more serious than other fractures.

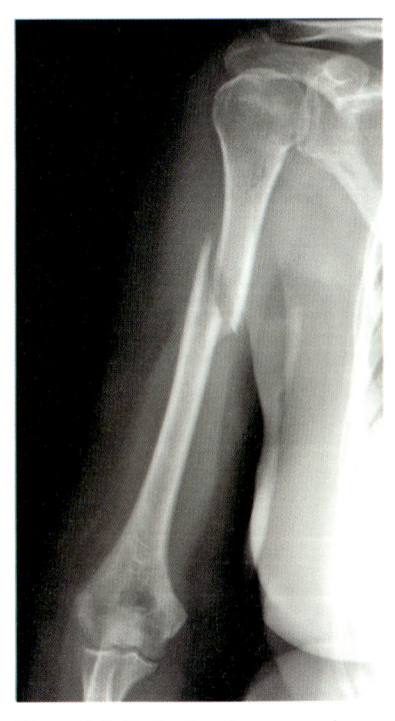

Figure 1.8.4b: Broken bones can be seen in an X-ray image.

Other problems of the skeletal system

8.4

From the age of approximately 35, the density of bones decreases naturally. In some people, the density drops below a healthy level and bones become fragile, making them prone to fractures. This condition is called **osteoporosis**. Treatment for osteoporosis includes taking drugs and calcium supplements to strengthen the bones.

Arthritis is a condition that affects the joints. In one form of arthritis, the cartilage at the end of the bones wears away and bones rub together. This can be very painful. In severe cases, the worn joint is replaced with an artificial joint.

> 4. Explain why sufferers of osteoporosis are prone to fractures.
> 5. Explain why arthritis can be so painful.

Did you know...?

On Earth, untreated osteoporosis can typically lead to a loss of 1.5% of bone mass per year. Astronauts can lose 1.5% of bone mass each month!

Medical advances

As technology improves, the diagnosis of fractures by X-rays has become more precise. Surgical techniques have also improved recovery from serious fractures.

As scientists learn more about osteoporosis, they can advise on how to avoid this disease. In the past, all that could be done was to treat the fractures.

As technologies have improved, joint replacements have become much more successful in improving movement and decreasing pain. However, operations such as hip and knee replacements carry risks of infection. There is also the possibility of decreased movement after the operation and the chance that the new joint can become displaced.

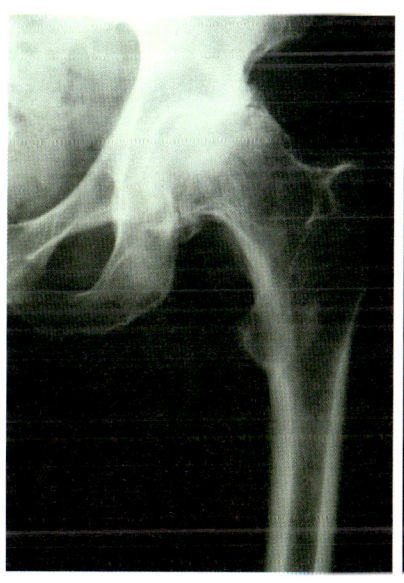

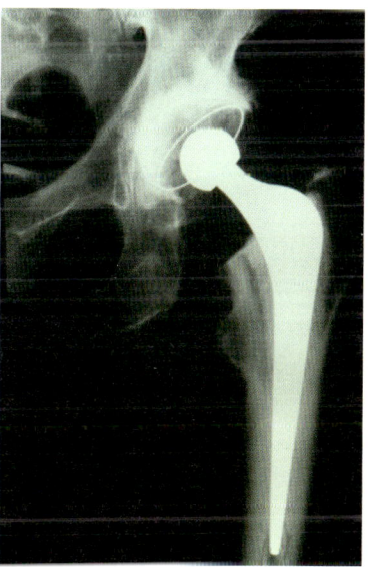

FIGURE 1.8.4c: X-ray images of a hip joint before and after the fitting of an artificial joint.

> 6. Suggest why improvements are likely to continue to be made in medical technology.
> 7. Explain why some arthritis patients decide not to have joint replacement surgery. Suggest what someone might say to persuade them that surgery would be a good idea.

Know this vocabulary

fracture
osteoporosis
arthritis

SEARCH: skeletal system problems   **183**

Organisms

Understanding organisation of organisms

We are learning how to:
- Define the terms tissues, organs and organ systems.
- Describe how some recreational drugs affect body systems.
- Suggest the effect of organ damage on other body systems.

The first simple multicellular organisms are thought to have evolved about 1.2 billion years ago. These eventually increased in organisation and size to form complex, multi-system, multicellular organisms. There are 15 different organ systems in a human, all working together to help us to survive. What can affect how well these systems work?

Cells, tissues and organs

Groups of similar body cells working together are called **tissues**. Examples of human tissues are muscle and bone. Different tissues work together to make up an **organ**. Every organ has a specific job – the eye is an organ made up of many different tissues including a lens and an iris. They work together to enable us to see. Examples of other organs are:
- the heart, which pumps blood to the cells;
- the kidneys, which cleanse the blood and balance water in the body;
- the brain, which allows us to control all parts of our body quickly.

Organs work together to make **organ systems**. An example of an organ system is the circulatory system.

1. Name three other organs and describe their functions.
2. The skin is described as an organ, not a tissue. Suggest why.

Effect of drugs on organs and systems

A drug is any substance that affects the way the body functions. Some drugs are taken to treat medical conditions, such as paracetamol for pain or anaesthetics before an operation. Drugs that are not used for medical reasons are **recreational drugs**.

Drugs can be grouped into four main groups:
- **Painkillers** relieve pain. Examples are paracetamol, codeine and morphine.
- **Stimulants** speed up body systems. Examples are caffeine, nicotine, cocaine, ecstasy and amphetamines.
- **Depressants** slow down body systems. Examples are alcohol, cannabis, tranquillisers (sleeping tablets) and heroin.
- **Hallucinogens** cause us to see things that do not exist. Examples are LSD and psilocybin mushrooms.

Each of these drugs affects the body in different ways. Table 1.8.5 summarises the organ systems that are affected by different drugs:

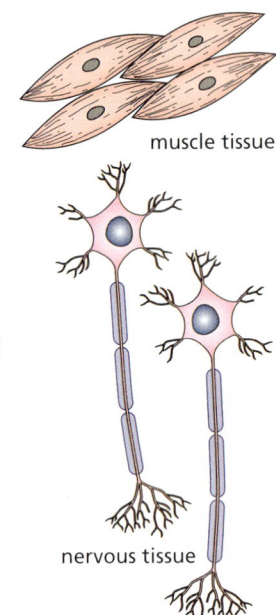

muscle tissue

nervous tissue

FIGURE 1.8.5a: Two types of tissue as seen under a microscope. What do you notice about the cells in each type of tissue?

Did you know...?

Each human consists of about 100 trillion (1×10^{12}) cells working together.

An effect on one organ system by a drug can have a knock on effect on other body systems.

8.5

TABLE 1.8.5: Which organ would paracetamol affect?

Type of drug	Effect on body	Organ system affected
painkiller	pain messages blocked; feelings of pain reduced or removed	nervous system
stimulant	increase in alertness and energy, brain activity increased, heart rate increased	nervous system, circulatory system
depressant	relaxed feelings or sleepiness, brain activity decreased, heart rate decreased	nervous system, circulatory system
hallucinogen	sense of reality distorted	nervous system

3. Which organ system do all drugs affect?
4. Ketamine is an animal tranquiliser used by vets for operations. Identify which group of drugs ketamine belongs to and describe which body systems the drug would affect.

Consequences of damage to organs

Examples of the organ systems in the human body are the **circulatory system**, the **respiratory system**, the **digestive system**, the **reproductive system**, the **immune system**, the **muscular skeletal system** and the **nervous system**. To work effectively, each of these systems relies on more than one organ working together, and on other systems. For example, the respiratory system relies on the lungs to collect oxygen from the air that we breathe (and to remove carbon dioxide from the body). It also relies on the circulatory system to pump the blood carrying the oxygen to all parts of the body (and to pump blood containing carbon dioxide back to the lungs to be breathed out). This oxygen is used to release energy in our bodies.

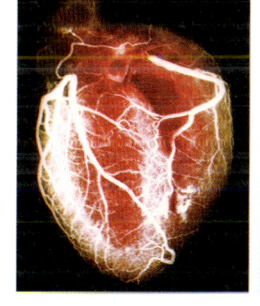

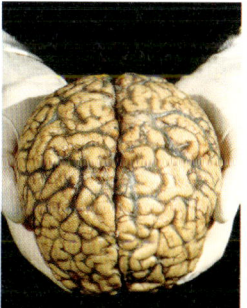

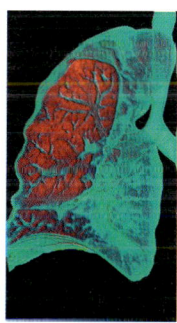

FIGURE 1.8.5b: What would be the consequences of damage to these organs?

If a person suffers from a heart condition, ineffective pumping of the heart can result in a lack of oxygen reaching all tissues of the body. So, even if the lungs are working well to bring air into the body and to absorb the oxygen that is needed, the person will feel tired and lack energy.

5. a) State the function of each of the organ systems listed in the text.
   b) Name an organ involved in the muscular skeletal system.
6. The nervous system relies on neurons to transmit impulses, carrying information around the body. It also relies on the brain to then interpret the information. Predict the consequences of damage to the brain on the nervous system.

Know this vocabulary

tissue
organ
organ system
recreational drugs
circulatory system
respiratory system
digestive system
reproductive system
immune system
muscular skeletal system
nervous system

SEARCH: organisation in multicellular organisms

Organisms

Describing animal and plant cells

We are learning how to:
- Describe the structures found in animal and plant cells.
- Explain the function of some of the structures within animal and plant cells.
- Communicate ideas about cells effectively using scientific terminology.

Every cell is a chemical processing factory, with hundreds of thousands of chemical reactions occurring every second. Without these reactions, the organism would die.

Cells as building blocks

All living organisms are made of **cells** – they are the building blocks of life. Cells cannot be seen except under a microscope. Some organisms are made of only one cell; most are made of millions of cells working together.

1. How can we see cells?
2. Is a cell living?

FIGURE 1.8.6a: An amoeba is a unicellular (single-celled) organism.

Common structures in animal and plant cells

All plant cells and animal cells have three main structures – the **nucleus**, the **cytoplasm** and the **cell membrane**.

Every cell, except red blood cells, contains a nucleus. The nucleus contains DNA, which controls the reactions inside the cell and is involved in making the cell reproduce.

The cytoplasm is a jelly-like material that makes up the bulk of the cell. All the chemical reactions occur here. Smaller structures within the cytoplasm, called organelles, make new materials to keep the cell and the organism alive.

The cell membrane surrounds the cell and contains the cytoplasm. The cell needs water, oxygen, glucose and nutrients – the membrane lets these in. During the chemical reactions, the cell makes waste products that it must get rid of, including carbon dioxide and urea. The membrane lets these substances out of the cell.

In the cytoplasm, special organelles called **mitochondria** convert glucose and oxygen into a form of energy that the cell can use.

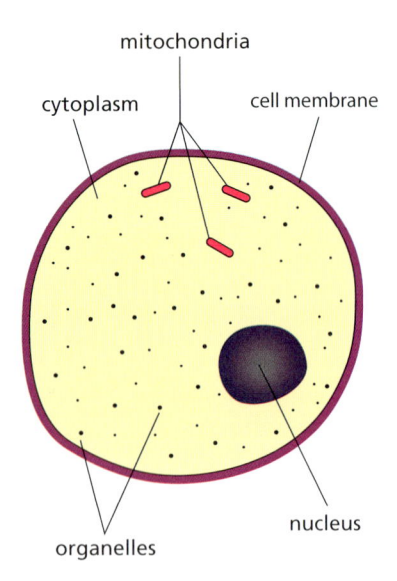

FIGURE 1.8.6b: The main structures of an animal cell.

186   AQA KS3 Science Student Book Part 1: Organisms – Movement *and* Cells

3. Which parts of the cell are found inside the cytoplasm?
4. What main substances can move through the cell membrane?

Differences between animal and plant cells

Animal cells are the simplest type of cell, containing a nucleus, cytoplasm, a cell membrane, and mitochondria in the cytoplasm. Plant cells share these parts, but also have other important structures.

The **cell wall** is an extra protective layer outside the cell membrane, made of cellulose. It gives the cell shape and strength.

The **vacuole** is a sac in the cytoplasm full of liquid, storing water, sugars, nutrients and salts. It provides internal pressure for the cell, keeping it firm and in shape. It also helps to control water movement inside the cell and between cells.

Leaf cells also contain small, round, green organelles called **chloroplasts**. These contain a green pigment called chlorophyll, which absorbs energy from the Sun and helps the plant make glucose.

Did you know...?

Many scientists believe a theory that mitochondria and chloroplasts evolved from bacterial cells. It is thought that large cells ingested the mitochondria and chloroplasts and they then evolved so that they could no longer exist outside a cell.

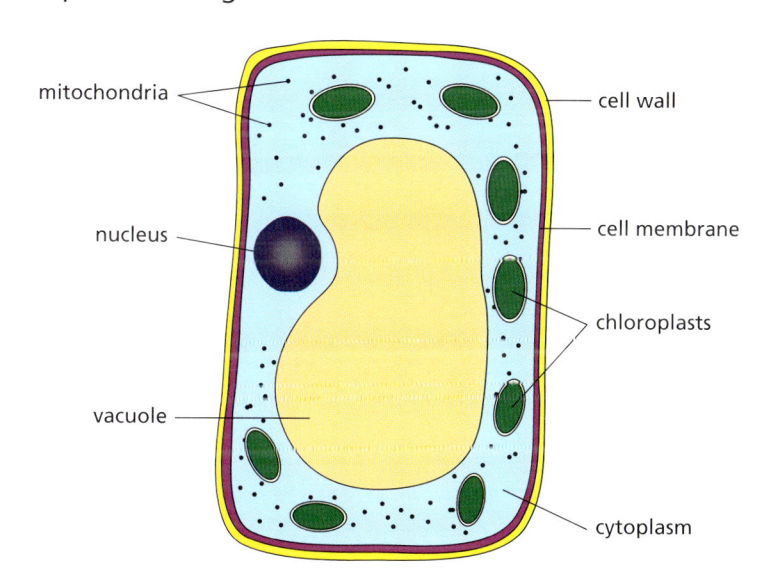

FIGURE 1.8.6c: A plant leaf cell.

5. Which two structures give a plant cell its shape?
6. Which cell do you think will be larger – a plant cell or an animal cell? Explain your answer.
7. Why do you think plant cells need extra structures that are not found in animal cells?

Know this vocabulary

cell
nucleus
cytoplasm
cell membrane
mitochondria
cell wall
vacuole
chloroplast

Organisms

Understanding adaptations of cells

We are learning how to:
- Recall the purpose of specialised cells.
- Identify examples of specialised plant and animal cells.
- Explain the structure and function of specialised cells.

All new cells in an organism start out exactly the same – these are called stem cells. When they grow, stem cells change their structure to carry out a certain job within the organism.

The right cells for the job

Many animal cells look very different from each other, although they contain the same basic structures. Cells become *specialised* so they can carry out a particular job. In an organism, many different jobs need to be done to keep it alive. These include movement, detecting information about the environment, sending impulses, carrying chemicals around the body, making chemicals the body needs, reproducing and absorbing food.

1. Where would you find cells that detect:
   a) light?   b) waves?   c) heat?
2. Explain why it is important that cells become specialised.

Specialised animal cells

Specialised cells have developed **structural adaptations** that enable the cell to carry out specific functions.

Nerve cells have very long extensions of cytoplasm. This enables them to carry messages from one part of the body to another.

Muscle cells are made from protein fibres that can rapidly expand and contract to create movement. They have the most mitochondria of all cells because they need lots of energy.

Sperm cells have tails and large heads. Their main job is to carry genetic material to an egg cell, so that it can be fertilised. Sperm cells have lots of mitochondria because they must swim long distances.

3. Name the types of animal cell in Figure 1.8.7a.

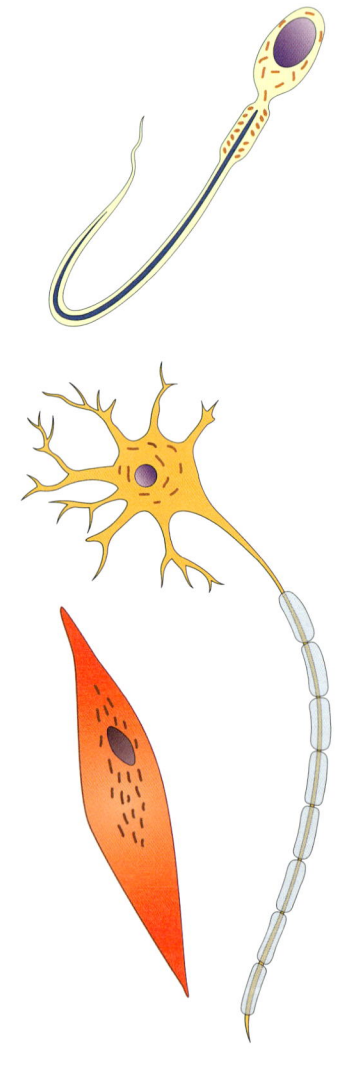

FIGURE 1.8.7a: Can you find the nucleus, cell membrane, cytoplasm and mitochondria of each of these specialised cells?

4. Which type of cell:
    a) transmits electrical messages?
    b) contracts and expands to create movement?
    c) carries genetic material for fertilisation?

Specialised plant cells

Plant cells are also highly specialised. Plants make their own food by a process called photosynthesis. Cells need to collect light and water, and take in carbon dioxide, and they produce a sugar, glucose. Many of the specialised cells in a plant are linked to this function.

Look at the leaf cell shown in Figure 1.8.6c. Chloroplasts trap sunlight needed for photosynthesis.

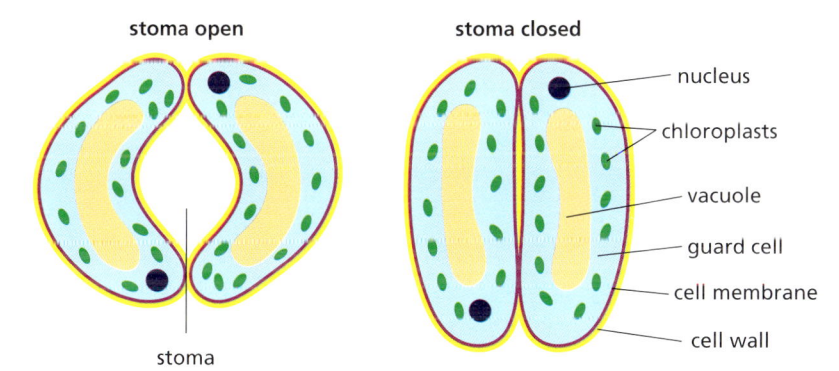

FIGURE 1.8.7b: Leaves have specialised cells called guard cells.

> **Did you know...?**
> There are more than 200 different types of specialised cells in the human body. In 2012, a Nobel Prize was awarded for the discovery that specialised cells can be changed to become stem cells.

Root hair cells (Figure 1.8.7c) have a long, thin extension called the root hair. This root hair provides a large surface area for water to be absorbed into the roots.

Other specialised cells in the leaf, called guard cells, are adapted to allow carbon dioxide into the leaf, Figure 1.8.7b. These cells can change shape to allow a space to open up so that the gas can move into the leaf.

5. Explain why it is important that plant cells are specialised.
6. Suggest why leaf cells contain lots of chloroplasts compared to other plant cells, such as root hair cells.
7. Look at Figure 1.8.6c showing a leaf cell and at Figure 1.8.7c showing a root hair cell. Describe the features of each and suggest how these features enable the cells to carry out their jobs.

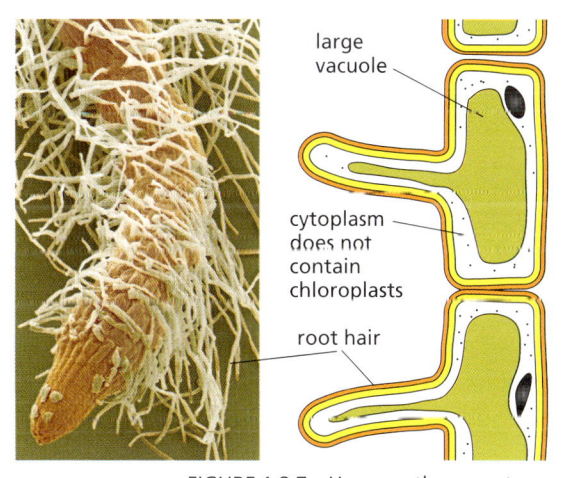

FIGURE 1.8.7c: How are these root hair cells different from the leaf cell shown in Figure 1.8.6c?

Know this vocabulary

specialised cell
structural adaptations

SEARCH: specialised cells 189

Organisms

Exploring cells

We are learning how to:
- Observe cells using a microscope and record findings.
- Explain how to use a microscope to identify and compare cells.
- Explain how developments in science can change ideas.

For many years, people believed that living things spontaneously appeared from non-living things. The invention of microscopes allowed scientists to observe cells and to understand how complex, but also structured, living things are.

Discovering cells

In 1590, the first **microscope** was developed and this allowed objects to be magnified. Robert Hooke developed this technique and, in 1667, he observed cells for the first time. From studying samples of cork bark, Robert Hooke discovered that organisms were made from simple building blocks, or cells. They are too small to be seen with the unaided eye.

1. State why we need a microscope to observe cells.
2. What is meant by 'magnified'?

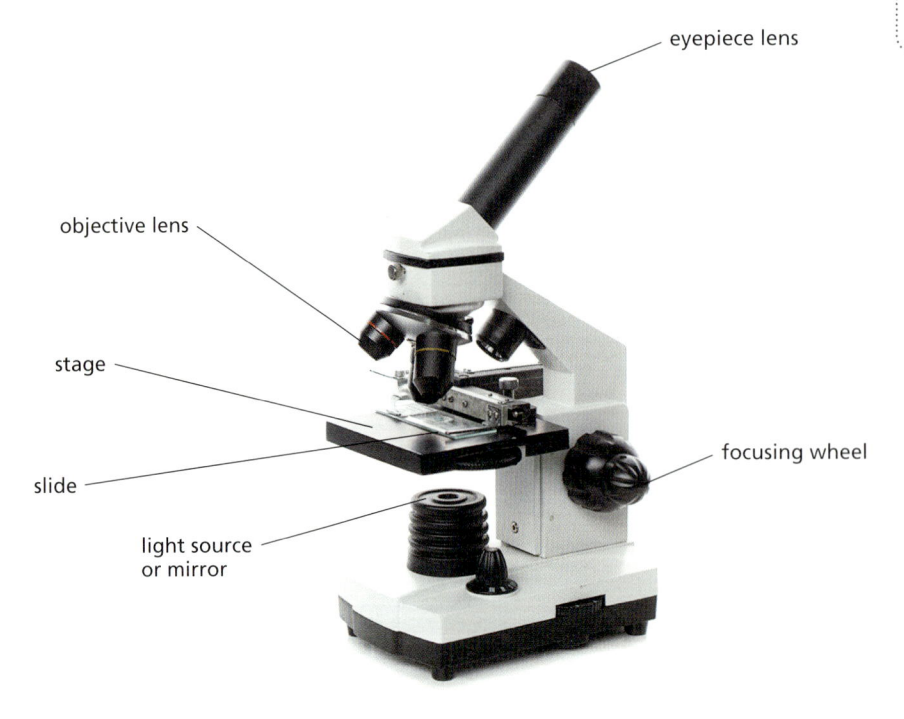

Did you know...?

Robert Hooke used the word 'cells' to describe what he saw under the microscope because he thought that the building blocks of the cork he was using looked like the cells of a monastery where monks lived.

FIGURE 1.8.8a: A light microscope.

190   AQA KS3 Science Student Book Part 1: Organisms – Movement and Cells

Observing cells

8.8

Figure 1.8.8a shows a light microscope. A specimen is placed on a glass slide on the stage. This is illuminated from beneath and two lenses (the objective lens and the eyepiece lens) then magnify the image. There is a selection of different objective lenses to allow a range of **magnifications**. The specimen is often stained with a dye to ensure that the features of the cells can be seen.

Cheek cells can be collected using a sterile cotton bud. These cells can then be rubbed onto a glass slide and methylene blue stain added. The slide is then placed on the stage and the objective lens with the lowest power used initially. The image is focused by gently moving the stage. Figure 1.8.8b shows an image of cheek cells viewed in this way. Figure 1.8.8c shows onion cells observed using iodine solution as a stain.

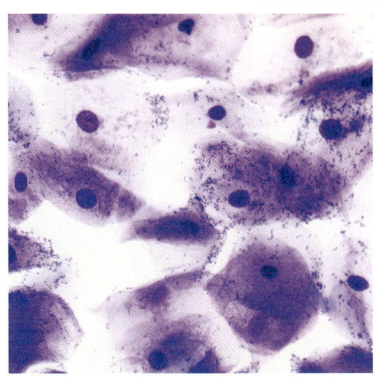

FIGURE 1.8.8b: Cheek cells stained with methylene blue viewed with a light microscope.

3. Explain why stain is often used on microscope specimens.
4. Describe what cheek cells and onion cells appear like under the microscope, and explain the function of the structures observed.
5. Compare the images in Figure 1.8.8b to the drawings of an animal and plant cell in Figures 1.8.6b and 1.8.6c.

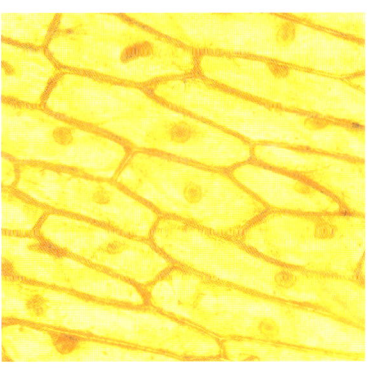

FIGURE 1.8.8c: Onion cells stained with iodine viewed with a light microscope.

Advances in observing cells

Microscopes have advanced hugely since Hooke's version in 1667. However, many are based on the same principles. As microscopes have improved, scientists are able to see smaller cells and smaller structures within cells. These advances have helped scientists to understand how cancer changes cells, for example. Understanding the changes caused is an important step in working out how to treat cancer.

Another type of microscope is the electron microscope. This uses a beam of electrons to form an image of a specimen. Electron microscopes allow greater magnification than light microscopes and have allowed scientists to see images of viruses (Figure 1.8.8d).

FIGURE 1.8.8d Image of a staphylococcus virus infection produced using an electron microscope.

6. Describe two differences between a light microscope and an electron microscope.
7. Suggest the importance of improving and developing microscope techniques and technologies.

Know this vocabulary

microscope
magnification

SEARCH: using a microscope

Organisms

Understanding unicellular organisms

We are learning how to:

- Recognise different types of unicellular organisms.
- Explain how unicellular organisms are adapted to carry out functions.
- Compare and contrast features of different unicellular organisms.

The oldest unicellular organisms were found in rocks dated to 3.8 billion years ago. They used chemicals in the ocean for 'food'. Around 3.5 billion years ago, unicellular organisms that could make their own food evolved. Unicellular organisms were the main form of life on the planet for nearly 2 billion years.

Unicellular organisms

Unicellular organisms are made up of just one cell. Each cell carries out all the life processes needed to exist independently. They differ from each other in their structure, how they feed and how they move. Algae are plant-like unicellular organisms containing chloroplasts and make their own food. Animal-like unicellular organisms, such as the amoeba (Figure 1.8.9a), take in food through their cell membrane. Some have developed tiny hairs to help them move, so they can find food or escape from predators. Some are themselves predators and will devour other unicellular organisms. Fungus-like unicellular organisms are called yeasts. They have a cell wall but cannot make their own food.

1. What is a unicellular organism?
2. Name three different unicellular organisms.
3. List three ways in which unicellular organisms differ from each other.

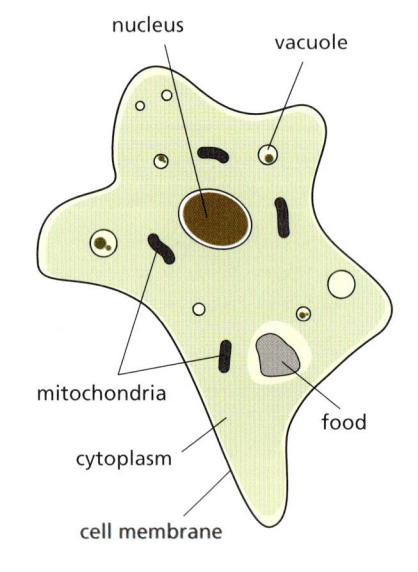

FIGURE 1.8.9a: An amoeba can carry out all life processes.

Prokaryotes

Unicellular organisms can be classified into two main groups – **prokaryotes** and **eukaryotes**. Prokaryotes are thought to have been the first organisms to live on Earth. They do not have a nucleus, and their genetic material floats within the cytoplasm. They have no, or few, organelles. **Bacteria** are examples of prokaryotes. They come in different shapes and sizes, live in different environments and have a range of food sources.

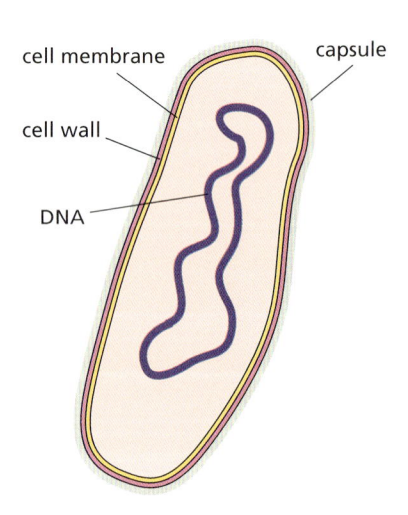

FIGURE 1.8.9b: A bacterium has no nucleus and no mitochondria.

Some bacteria take in chemicals from their environment, such as iron and sulfur, and use these as food. Others contain chlorophyll and use sunlight to make their own food. Some can absorb nutrients through their cell membrane by **diffusion**. Bacteria can be found in extreme conditions, from under-sea volcano vents to places with temperatures well below freezing.

8.9

4. Look at Figure 1.8.9a and Figure 1.8.9b. Which is a prokaryote and which is a eukaryote? Explain your answer.

5. Describe how some prokaryotes are adapted to:
   a) carry out photosynthesis;
   b) absorb nutrients from the environment.

Did you know...?

Nummulites are fossils of the largest known unicellular organisms. Nummulite fossils as large as 16 cm across have been found. Some are thought to have lived for over 100 years.

Eukaryotes

Eukaryotes contain a nucleus, surrounded by a nuclear membrane. They also contain many organelles, including mitochondria, chloroplasts and vacuoles.

Examples of eukaryotes are euglena (a type of alga containing chloroplasts), yeast, amoeba, and paramecium.

Eukaryotes can be up to 200 times bigger than prokaryotes and often have external features to help them to survive. The amoeba can move around because its cytoplasm can flow; the paramecium has cilia that beat and enable it to move, and the euglena has a flagellum, or tail, to enable it to move.

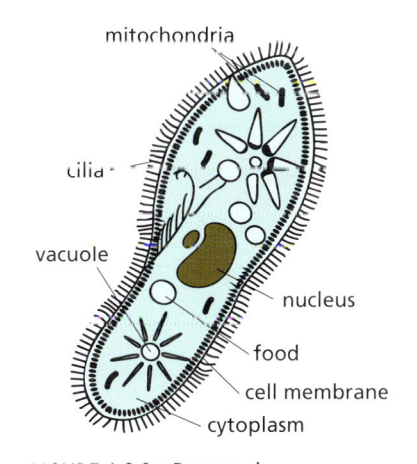

FIGURE 1.8.9c: Paramecium.

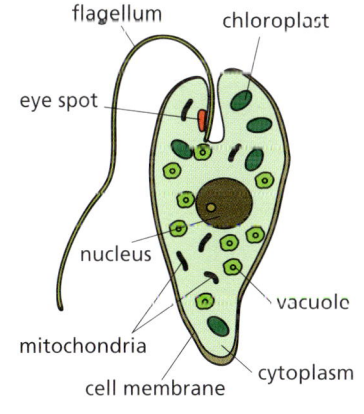

FIGURE 1.8.9d: Euglena.

6. Look at Figure 1.8.9d. How does the euglena get its food?

7. Describe three different ways in which unicellular organisms move.

8. Summarise, in a table, the main similarities and differences between unicellular organisms – compare prokaryotes, paramecium (eukaryote) and euglena (eukaryote).

Know this vocabulary

unicellular
prokaryote
eukaryote
bacteria
diffusion

SEARCH: unicellular organisms

Organisms

Checking your progress

To make good progress in understanding science you need to focus on these ideas and skills.

Identify the main bones of the skeleton.	Describe the functions of the skeleton.	Explain how different parts of the skeleton are adapted to carry out particular functions.
Describe the role of skeletal joints.	Identify some different joints and explain the role of tendons and ligaments in joints.	Compare the movement allowed at different joints and explain why different types of joints are needed.
Recall that muscles contract to move bones at joints.	Identify muscles that contract to cause specific movements.	Explain how muscles work antagonistically to bring about movement, and evaluate a model.
Recognise and label basic and specialised animal cells and plant cells.	Describe the functions of the nucleus, cell membrane, mitochondria, cytoplasm, cell wall, vacuole and chloroplast.	Compare and contrast the similarities and differences between specialised animal cells and plant cells.
Describe unicellular organisms – including yeast, bacteria, euglena, paramecium and amoeba – as being either prokaryotes or eukaryotes.	Describe the function of specialised parts of different unicellular organisms.	Explain how different structures help organisms to survive.

8.10

- [ ] Put the terms cell, tissue, organ and organ system in order of hierarchy, naming some common tissues, organs and organ systems in humans.

- [ ] Explain the terms cell, tissue, organ and organ system and the function of some of the main organ systems in the body.

- [ ] Explain the relationship between different organs of the body and predict the consequences of damage to specific organs.

- [ ] Recall that a microscope magnifies an image and allows us to see objects not visible to the naked eye.

- [ ] Describe and demonstrate how to observe animal and plant cells under the microscope and explain observations.

- [ ] Explain the importance of the development of microscopy techniques, using examples.

Organisms

Questions

KNOW. Questions 1–7

See how well you have understood the ideas in this chapter.

1. Identify the femur in Figure 1.8.11a. [1]
2. What are the small bones that make up the backbone called? [1]
   - a) ligaments
   - b) joints
   - c) vertebrae
   - d) tendons
3. How does the ribcage protect the lungs? [2]
4. Explain how muscles cause bones to move. [2]
5. Which of the following is a unicellular organism? [1]
   - a) nerve cell
   - b) cytoplasm
   - c) amoeba
   - d) flowering plant
6. Where in the cell would the most diffusion take place? [1]
   - a) nucleus
   - b) cell membrane
   - c) chloroplast
   - d) mitochondria
7. Explain why stain is often used on cell samples before observing with a light microscope, and give an example. [2]

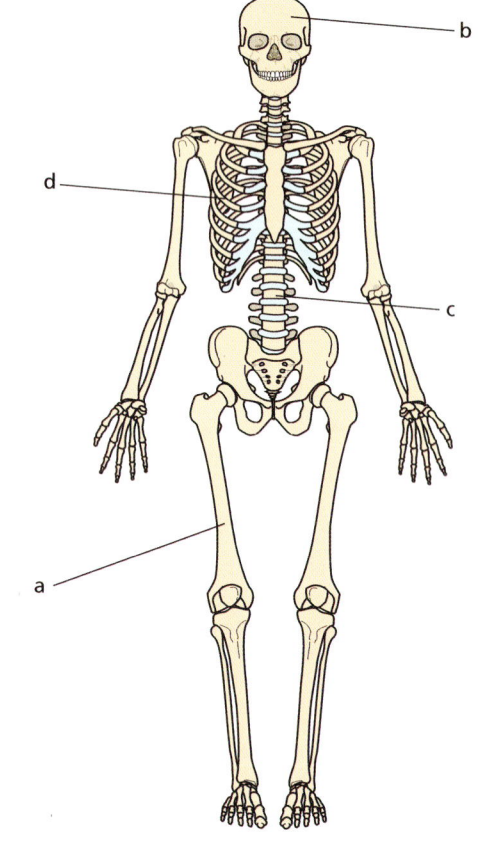

FIGURE 1.8.11a

APPLY. Questions 8–12

See how well you can apply the ideas in this chapter to new situations.

8. When a muscle underneath the toe contracts to move the toe down, its antagonistic muscle is: [1]
   - a) contracting;
   - b) relaxing;
   - c) neither contracting nor relaxing;
   - d) pushing.
9. Describe how movement at the elbow joint would be different if it were a ball and socket joint. Explain your answer. [2]
10. A sample of cells is observed under a microscope. Each cell has a cell wall, cell membrane and nucleus, but no chloroplasts. Suggest whether this cell is from a plant or an animal, giving reasons. Which part of the organism might these cells have been taken from? [4]

8.11

11. Some plants live in conditions of low light on the floor of thick forest. Which of the following features are likely to help them to survive? [1]

    a) They will have brightly coloured petals.
    b) Their leaves will be dark green, packed with more chloroplasts than ordinary leaves.
    c) They will have fewer root hair cells.
    d) Their seeds will have small mass so they can be blown by the wind.

12. Some scientists discover a new unicellular organism. What features would enable them to classify it as an alga? [2]

EXTEND. Questions 13–16

See how well you can understand and explain new ideas and evidence.

13. A bodybuilder has strained his tricep muscles and has been advised to rest his arm. He asks if he could carry on using hand weights to build up his biceps while still resting his triceps. Explain why this is not possible. [4]

14. Imagine that a strange skeleton of an unknown animal is found at an archaeological dig. The backbone of the skeleton is one long bone. Suggest what this tells us about movement of the animal, compared to the movement of humans. [2]

15. Look at Figures 1.8.11b and 1.8.11c. Identify which is from a light microscope and which is from an electron microscope. Explain your answer. [2]

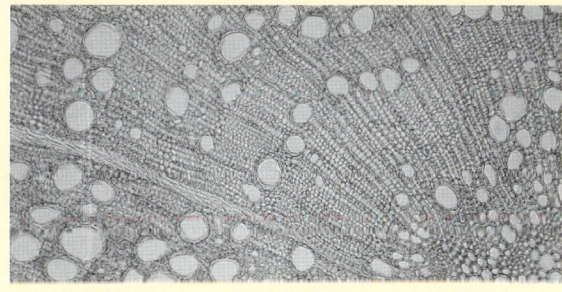

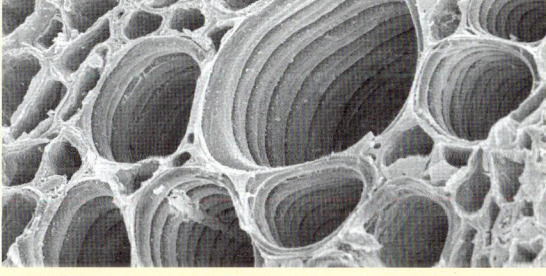

FIGURE 1.8.11b

FIGURE 1.8.11c

16. Table 1.8.11 shows the surface area and volume measured for three different unicellular organisms. Suggest which of the organisms is likely to be more efficient at absorbing nutrients from its environment and explain why. [2]

TABLE 1.8.11

Organism	Surface area (cm^2)	Volume (cm^3)	SA/volume ratio
A	12	6	2:1
B	10	2	5:1
C	12	4	3:1

Ecosystems
Interdependence *and* Plant reproduction

Ideas you have met before

The environment

All living things depend on one another to survive.

A food chain shows how each living thing gets food for energy.

fox

rabbit

grass

Reproduction in plants

The roots, stems, leaves and flowers of a plant each have a specific purpose.

Flowers enable reproduction in plants, through pollination and seed dispersal.

Plants have evolved different ways of carrying out these processes.

9.0

In this chapter you will find out

Relationships in the environment

- In any environment there are many interlinked food chains. These can be disrupted by factors such as toxins entering the food chain, or disease.
- Food chains usually start with a plant or plant material, called a producer.
- Animals that eat plants and other animals are consumers and these are found at different levels of a food chain.
- The availability of food is crucial, and insects can play an important role in food security.

How plants are adapted to reproduce

- Flowers are adapted in many ways to attract pollinators or use the wind to help pollination.
- A pollen grain contains the male sex cell in plant reproduction and the ovule is the female sex cell; fertilisation is the meeting of these two cells.
- Plants have evolved different mechanisms to disperse their seeds, increasing their chances of survival.
- We can use models to investigate the efficiency of seed dispersal.

Ecosystems

Understanding food webs

We are learning how to:
- Describe how food webs are made up of a number of food chains.
- Make predictions about factors affecting plant and animal populations.
- Analyse and evaluate changes in a food web.

Food chains show the feeding relationships between living organisms. If something happens to disrupt part of the chain, it can have serious knock-on effects through the whole chain.

The ups and downs of food chains

The organisms in a **food chain** are dependent on each other. For example, in Figure 1.9.1a, grass is eaten by rabbits, which in turn are hunted and eaten by foxes. The grass captures the energy from sunlight to photosynthesise and make glucose. The glucose provides energy for the plant to grow. When a rabbit eats grass, some of the energy left in the grass is transferred to the rabbit. The rabbit uses some of this energy to move and grow. When a fox eats a rabbit, the remaining energy in the rabbit is transferred to the fox.

Changes in the number of one organism in an area – its **population** – affect other organisms in the same food chain.

- The number of plants in an area can be affected by the amount of rain, sunlight, minerals and space available to grow.
- The number of animals can be affected by the availability of food, habitats, mates and water, and by disease.

Look at Figure 1.9.1a again and then answer these questions.

1. What would happen to the numbers of rabbits and foxes if all the grass died out?
2. What would happen to the amount of grass and foxes if all the rabbits died out?
3. Why is it a good idea for an organism to have different sources of food?

Food webs and trophic levels

Most animals eat many different things and are involved in many different food chains. These food chains can be linked together in a **food web**, which shows how the food chains are connected. Food webs can be complex.

FIGURE 1.9.1a: A simple food chain.

Did you know…?

Fungi, such as mushrooms, are very important as decomposers in food webs. Each fungus has chemicals, called enzymes, that decompose only a small number of specific materials. Some fungi can even decompose jet fuel!

FIGURE 1.9.1b: Shaggy ink cap fungus (*Coprinus comatus*).

In a food web:
- **producers** make their own food;
- **consumers** eat other organisms, either plants, animals or both;
- **decomposers** break down dead plant and animal material; the nutrients released are recycled in soil or water.

These rankings are called **trophic levels**. The trophic level of an organism is the position it occupies in a food chain.

> 4. Identify a consumer in the food web in Figure 1.9.1c. What does this consumer eat?
> 5. Describe the role of a decomposer and give an example from the food web in Figure 1.9.1c.
> 6. If all the mice died, what could happen to the rabbits in the food web?

Knock-on effects

FIGURE 1.9.1c: A simple food web. What do the arrows in the food web mean?

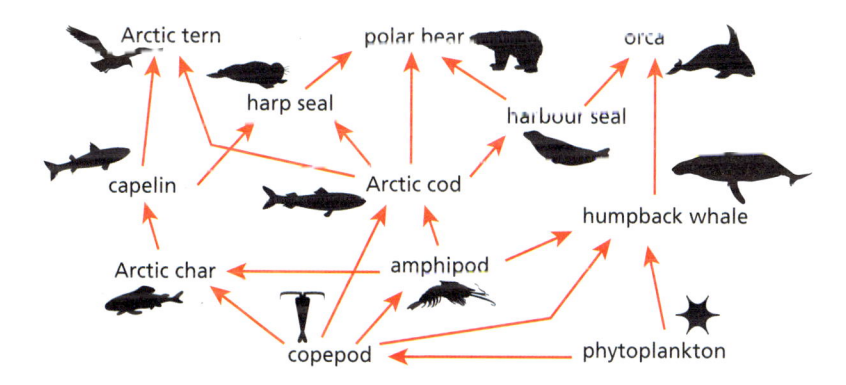

FIGURE 1.9.1d: An Arctic food web.

Look at Figure 1.9.1d. Harbour seals, harp seals and Arctic terns all feed on Arctic cod. If the Arctic cod catch a disease and die, the Arctic terns and harp seals will eat more of their other prey. The harbour seal only eats Arctic cod, so they will die too. There will be no cod to eat amphipods and copepods, so there will be more food for the humpback whale and Arctic char, and their populations will increase.

> 7. Harp seal populations are controlled by killing them – this is called 'culling'. Analyse and evaluate the impact of culling the majority of harp seals.
> 8. Explain how this food web shows that energy is transferred from the copepod to the Arctic tern.

Know this vocabulary

food chain
population
food web
producer
consumer
decomposer
trophic level

SEARCH: food webs

Ecosystems

Understanding the effects of toxins in the environment

We are learning how to:
- Describe how toxins pass along the food chain.
- Explain how toxins enter and accumulate in food chains.
- Evaluate the advantages and disadvantages of using pesticides.

Otters nearly became extinct in the south of England in the 1960s. What caused this? Why were otters more affected than other animals? Why do we use chemicals in agriculture?

Why are chemicals used in agriculture?

In recent times, the global human population has increased dramatically. Food needs to be grown more quickly to feed the growing number of people. Soils are quickly depleted of the nutrients needed to grow healthy crops. Nowadays, farmers rarely mix keeping animals with growing crops, so they do not have the supplies of cattle manure to replace the nutrients naturally. Instead, artificial **fertilisers** and nutrients are used to replenish the soil.

Insecticides and **pesticides** are chemicals used to kill insect pests and other small creatures that damage crops.

1. Why do farmers use chemicals in agriculture?
2. Why do most modern farms not use manure on their fields?

FIGURE 1.9.2a: Why is the farmer adding artificial fertiliser, not manure?

Chemicals entering the food chain

Toxins can enter the food chain in several ways.

- Fertilisers dissolve in water and are washed off the fields by rain into rivers and reservoirs.
- Chemicals used by farmers to kill weeds or insects contaminate small creatures that are eaten, or the chemicals are washed or blown into waterways.
- Water runs off urban streets into waterways.
- Soft mud acts like a sponge that slowly soaks up the toxins. Plants absorb these through their roots.
- Some chemicals fall from the air, such as mercury released by coal-burning power plants.

FIGURE 1.9.2b: Insects covered in insecticide are eaten by other animals.

A consumer may eat the plants containing the toxins; other (secondary) consumers eat that (primary) consumer; and so on up the food chain.

> **3.** Give examples of a primary consumer and a secondary consumer that could be affected by pesticides used on farmland.
>
> **4.** Explain how toxins enter the food chain.

Accumulation of toxins in the food chain

Organisms at the start of a food chain can take up small amounts of toxins. The higher up the organism is in the food chain, the more concentrated the toxin will become – eventually it is so concentrated that it can kill the top predator.

A pesticide called DDT was used in the 1960s. It killed insects that were damaging crops, but it ran off into rivers and contaminated plants. The small animals and fish further up the food chain collected more and more of the toxin because it stayed in their bodies. This process is called **bioaccumulation**.

Otters that ate the fish were killed and almost became extinct in the south of England.

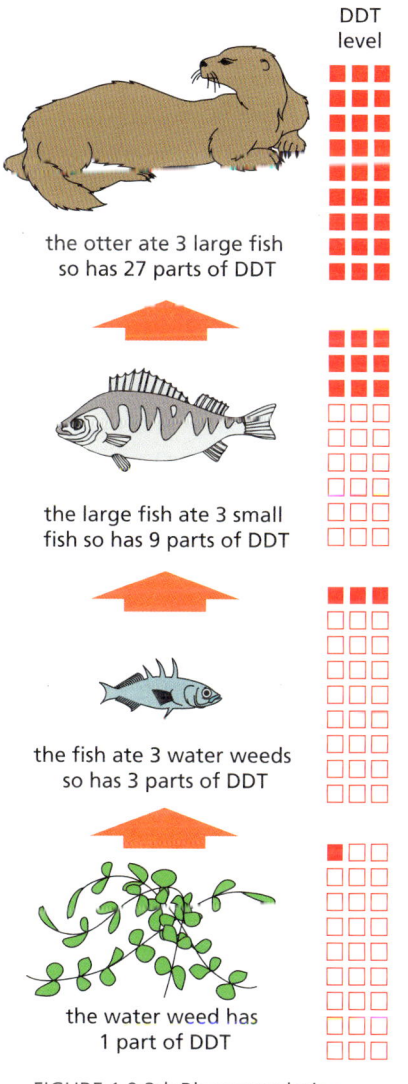

the otter ate 3 large fish so has 27 parts of DDT

the large fish ate 3 small fish so has 9 parts of DDT

the fish ate 3 water weeds so has 3 parts of DDT

the water weed has 1 part of DDT

FIGURE 1.9.2d: Bioaccumulation of DDT in a food chain.

FIGURE 1.9.2c: How does mercury released into the air get from here into an organism?

Did you know...?

DDT was banned worldwide in 2001. The only remaining legal use of DDT is to control malaria-carrying mosquitoes. Modern insecticides do not accumulate in food chains.

Know this vocabulary

fertiliser
insecticide
pesticide
toxin
bioaccumulation

> **5.** Explain why otters were in danger of extinction because of DDT.
>
> **6.** If the otter population declined, how would this affect the river ecosystem?

SEARCH: bioaccumulation in food chains

Ecosystems

Exploring the importance of insects

We are learning how to:
- Describe the role of pollination in crop production.
- Explain why artificial pollination is used for some crops.
- Evaluate the risks of monoculture on world food security.

Food security refers to the availability of food and the ability to obtain it. What is the role of bees and other insects in our food security? How does agricultural practice impact on food security?

Fruit production and bees

Bees are vital in pollinating fruit crops. **Pollination** is successful when flowers receive healthy pollen at the best time. The better the pollination of apples and pears, the more numerous and larger the fruits.

Anything that interferes with bee activity, such as disease or adverse weather, will reduce pollination. Bee colony numbers in Britain have fallen dramatically. The reduced pollination has lowered fruit yields and hence the earnings of fruit growers – the apple harvest in 2012 was 50% lower than expected. This resulted in a higher cost of apples in the shops.

Recent research has found that the fall in wild bee populations, caused by habitat destruction, is having a greater impact than the fall in honeybee numbers. This is because wild bees are twice as effective as honeybees in pollinating orchards.

FIGURE 1.9.3a: Honeybee hives are placed in orchards to ensure pollination.

1. Why do fruit growers put beehives in their orchards?
2. How can we help wild bee colonies to survive and grow?

Ensuring pollination

In south-west China, wild bees have become extinct because of overuse of pesticides and the destruction of their natural habitats. Apple and pear farmers now hand-pollinate their trees, using pots of pollen and paintbrushes to pollinate each flower individually.

Crops of cucumbers, tomatoes and peppers are also often hand-pollinated. Date palms have male and female plants; natural pollination therefore requires trees of both types.

FIGURE 1.9.3b: These women are hand-pollinating blossom on pepper plants.

By using hand-pollination, date farmers need only grow female trees and so avoid wasting space by growing male plants.

There are not enough humans in the world to pollinate all of our crops by hand. In addition, hand-pollinated fruits are often smaller than those pollinated by bees. Scientists are trying to develop a robotic bee that could be used to pollinate plants artificially and support the work of real bees.

> 3. Why is artificial pollination vital to fruit growers in China?
>
> 4. What are the advantages and disadvantages of artificial pollination?

Tackling food security

Evidence from around the world shows that yields of insect-pollinated crops are falling and are becoming ever more unpredictable. This is especially true in the areas with the most intensive farming. Where single crops are grown in vast fields – a practice called **monoculture** – there are not enough insects to go around.

Almond orchards cover hundreds of square miles in California. Bees cannot survive naturally in these areas because the flowering time is too short and there are no other plants for them to feed on.

FIGURE 1.9.3c: Monoculture is a modern agricultural practice that destroys the pollinators' natural habitats.

Some poor countries use monoculture to grow huge quantities of crops that they sell to richer countries, such as coffee, cocoa and bananas. Little fertile land is left to grow food crops for the local people, who then suffer food insecurity.

> 5. Evaluate the practice of monoculture in agriculture.
>
> 6. Suggest how farmers can ensure pollination in monocultural systems.

Did you know…?

It has been suggested that 'travel stress' caused by bees being shipped from pollination site to pollination site is partly to blame for disorders in bee colonies that hugely reduce their population.

Know this vocabulary

food security
pollination
monoculture

SEARCH: the importance of insects

Ecosystems

Exploring ecological balance

We are learning how to:
- Describe ways in which organisms affect their environment.
- Explain why prey populations affect predator populations.
- Evaluate a model of predator–prey populations.

Organisms are not isolated in their environment. They interact with other individuals of their own species, with other species and with their physical environment. The study of the interactions between organisms and their environment is called **ecology**. In what ways do organisms interact? How does one organism affect others?

How organisms affect the environment

All organisms cause changes in the **environment** where they live. An organism's behaviour depends on the nature of its environment. This includes factors such as:
- the types and numbers of other organisms present;
- the availability of food and resources;
- physical characteristics of the environment.

The living and non-living things in an area make up the **ecosystem**. Cattle that stay in one place for a long time will overgraze and destroy the plant life. Without plants to hold it, topsoil runs off into streams and lakes, causing habitat loss for organisms living in the fields. Soil fills the bottom of the streams and lakes and absorbs water, causing a drop in the water volume. This can affect the range and number of plants and animals living in the water. This then affects the range and number of organisms that can feed from plants and animals in the streams and lakes. This reliance of organisms on one another is known as **interdependence**.

> 1. What is 'ecology'?
> 2. Describe examples of how organisms affect their environment.

Competition

Competition is an example of interdependence. Organisms in an ecosystem are continually competing for the same limited resources. For example, if there is only a small number of fruit trees and one animal is taller and, therefore, better at reaching the fruit than a shorter animal, then the shorter animal is less likely to feed and may starve.

FIGURE 1.9.4a: How can cows affect grass when they graze?

Did you know…?

Big cats are examples of predators adapted for efficient hunting. One of the cheetah's best hunting skills is its ability to run at high speed. It can run faster than any other land animal, accelerating from 0 to 100 km/h (62 mph) in about three seconds.

FIGURE 1.9.4b: A cheetah.

Some organisms co-exist by specialising – for example, different plant roots may access water at different depths in the soil.

3. Explain how specialisation can help to reduce competition.

Predators and prey

The relationship between **predator** and **prey** is probably the most important form of interdependence. Predators need to be adapted for efficient hunting to catch enough food to survive. Prey species must be well adapted to escape their predators to ensure their survival. If the prey population grows, predator numbers will respond to the increased food supply and grow too. Increased predator numbers will reduce the food supply so that it can no longer supply the predator population. The resulting pattern of prey and predator numbers is shown in Figure 1.9.4c.

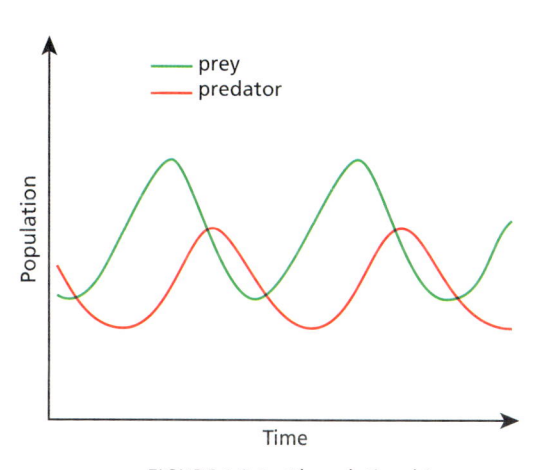

FIGURE 1.9.4c: The relationship between predator and prey numbers.

The effects of predator and prey, such as lion and antelope, on each other can be explored using a model. Make cards labelled 'lion' or 'antelope'. You will toss these on the table, which acts as the habitat of the animals. When a lion card lands on an antelope card, this models the lion catching and eating the antelope. The antelope 'dies' (and is removed from the game). When the lion card does not land on *three* antelope cards, this models the lion having insufficient food and it 'dies' (and is removed from the game).

Start by tossing three antelope cards and one lion card. Remove any cards of animals that would die. Record the outcome for this generation (1). Next, double the number of remaining antelope to model them reproducing, and repeat using one lion. Again, record the outcome (generation 2). Continue doubling the number of antelope left at the end of each generation. If a lion survives (by catching at least three antelope), it then reproduces so introduce a second lion in the next generation. Continue for approximately 16 generations and record the outcome of each.

FIGURE 1.9.4d: Ladybirds are used to control aphid populations.

4. Draw a graph to show predator and prey numbers at each generation of your model.

5. Predict the shape of the graph for generations 17 to 25.

6. Is this a good model? Explain your answer.

7. What variables affect the numbers of predators and prey in a population?

8. Explain how prey populations affect predator populations.

Know this vocabulary

ecology
environment
ecosystem
interdependence
competition
predator
prey

SEARCH: predator–prey relationships

Ecosystems

Exploring flowering plants

We are learning how to:
- Identify parts of flowering plants.
- Describe the function of the parts of flowering plants and link structure with function.
- Evaluate the differences between wind-pollinated plants and insect-pollinated plants.

The first plants on Earth were mosses. These relied on moisture and touch to transfer pollen. The first flowering plants, using wind and insects to transfer pollen, are thought to have evolved about 200 million years ago. Nowadays about 70 per cent of plant species use insects, birds or mammals to transport pollen.

Flowers as reproductive organs

Most flowers have male and female parts. The male part is the **stamen**, consisting of an **anther** and a **filament**. The anther produces **pollen**, which contains the male sex cell.

The female part is the **carpel**. This consists of an **ovary** (with the female sex cells in the ovules), the **style** and the **stigma**, which has a sticky top.

The purpose of the flower is to produce pollen in the anther and transfer it to the stigma of a different flower. This process of transferring pollen is called **pollination** and is mainly achieved using wind, insects, birds or bats.

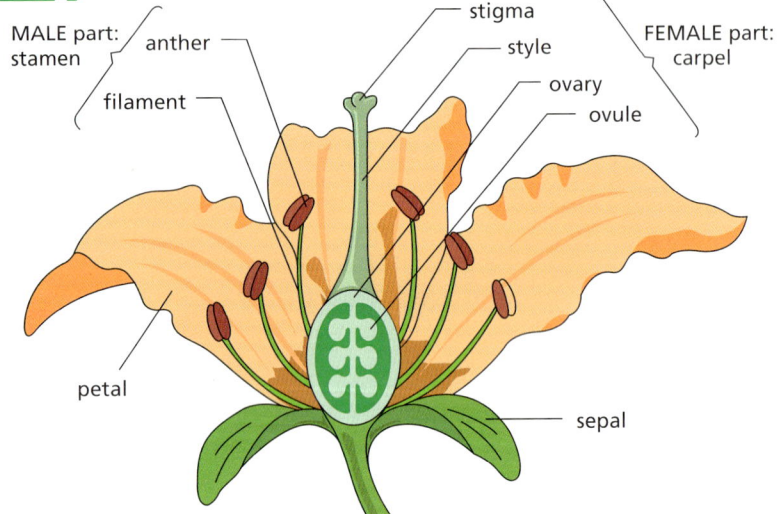

FIGURE 1.9.5a: Male and female parts of a flower.

1. Identify the following parts of the flower in Figure 1.9.5b: anther, filament, stamen, stigma, style, ovary.
2. What differences can you see between the two flowers in Figure 1.9.5d?

FIGURE 1.9.5b: A flowering Tulip.

208   AQA KS3 Science Student Book Part 1: Ecosystems – Interdependence *and* Plant reproduction

Attracting insects

Most insect-pollinated plants produce brightly coloured flowers with sweet smells to attract insects. Many also produce nectar deep inside the flower. This is a sugary fluid that encourages insects into the flower. Pollinators such as bees collect the pollen and nectar as food sources. Plants produce a lot of pollen to increase the chances of successful pollination.

3. Describe different ways plants encourage insects to visit them.
4. Why do plants use such a diverse range of methods of attracting pollinators?

Wind or insect pollination?

There is no guarantee that the wind will successfully transfer the pollen from one plant to the stigma of another plant, so wind-pollinated plants produce millions of pollen grains to improve their chance of success. Some stigmas evolved to become large and feathery so as to capture pollen floating on the wind. Even so, there is no guarantee that the pollen from the same species will land on the plants.

Insect-pollinated plants produce far less pollen than wind-pollinated plants, but use other mechanisms to attract insects. However, some insects eat parts of the flower and plant, so flowers have developed mechanisms to avoid this, such as producing toxins and growing spikes.

FIGURE 1.9.5c: This orchid mimics the appearance of a female wasp. The male wasp visits the flower and becomes covered in pollen.

Did you know...?

The oldest known pollen grains were found on the bodies of tiny insects encased in amber. The pollen was thought to be over 200 million years old. Fossilised pollen has provided evidence of how plant life on Earth has evolved.

Know this vocabulary

stamen
anther
filament
pollen
carpel
ovary
style
stigma
pollination

FIGURE 1.9.5d: Which flower is wind pollinated and which is insect pollinated?

5. Discuss the advantages and disadvantages of wind pollination and insect pollination.

SEARCH: wind and insect pollinated flowers

Ecosystems

Exploring fertilisation

We are learning how to:
- Describe the process of fertilisation in plants.
- Describe the role of pollen tubes.
- Explain how seeds are formed.

The world's chocolate supply depends on midges. These tiny flies are the only insects that can pollinate the cacao plant. Once fertilised, the plant produces seeds, which are used to make chocolate.

From pollination to fertilisation

Pollination ends when pollen from one flower reaches the stigma of another flower. A pollen grain contains the male sex cell. The female sex cell, the ovule, is found in the ovary. For **fertilisation** to occur, the ovule and the pollen cell must meet.

- As the pollen sits on the stigma, a **pollen tube** grows out of the pollen grain through the stigma and style and down into the ovary.
- The nucleus of the pollen cell travels down the tube into the ovary.
- The nucleus of the pollen cell meets the nucleus of the ovule; this is fertilisation.
- This fertilised ovule will eventually develop into a new plant.

1. Describe how pollination and fertilisation differ.

Pollen tubes

When a stigma is ripe for fertilisation, it secretes a sugary fluid onto its surface. It is this sugar that stimulates the growth of pollen tubes. The sugar provides the energy needed for the tubes to grow.

The concentration of sugar affects how well pollen tubes grow. This can be investigated by adding sugar solutions of different concentrations to the stigmas of plants and measuring the pollen tubes. Some results are shown in Table 1.9.6.

FIGURE 1.9.6a: Seed pods on a cacao plant.

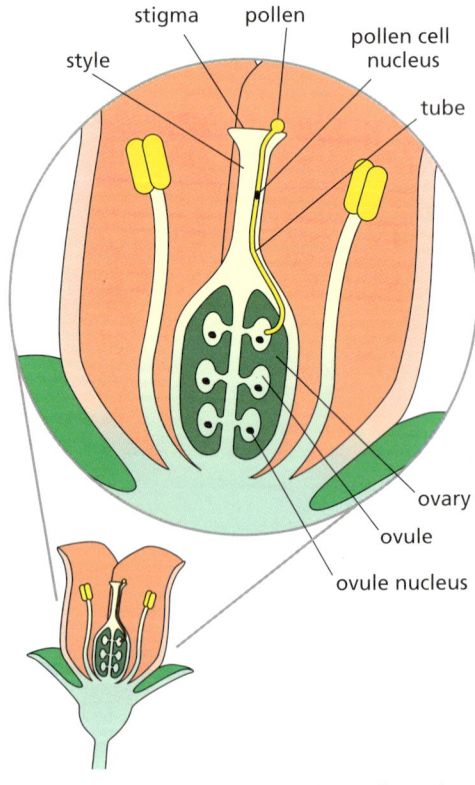

FIGURE 1.9.6b: How do you think a pollen tube is formed – from one cell or many?

TABLE 1.9.6: The effect of sugar on the growth of pollen tubes.

Sugar concentration (%)	5	10	15	20
Growth of pollen tubes (micrometres, µm)	250	350	450	200

2. Plot a graph of the data in Table 1.9.6.
3. Describe the pattern shown by the data.
4. Suggest what you would need to control if you were investigating how sugar concentration affects the growth of pollen tubes.

Development of seeds

Following fertilisation, many of the parts of the flower fall off because they are no longer needed. This includes the petals, sepals and stamens. Each fertilised ovule then becomes a **seed**. The outer layer of the seed becomes hard and the seed dries out (imagine an apple seed and how it feels). The ovary develops into a **fruit**. The fruit protects the seeds until they are ripe and ready to form a new plant.

In science, a fruit is defined as an ovary after fertilisation, containing seeds.

Did you know...?

Pollen that does not land on a stigma remains in the environment. It is the primary cause of hay fever and allergies. Pollen counts are made by counting how much pollen lands on a greasy spinning rod over a 24-hour period.

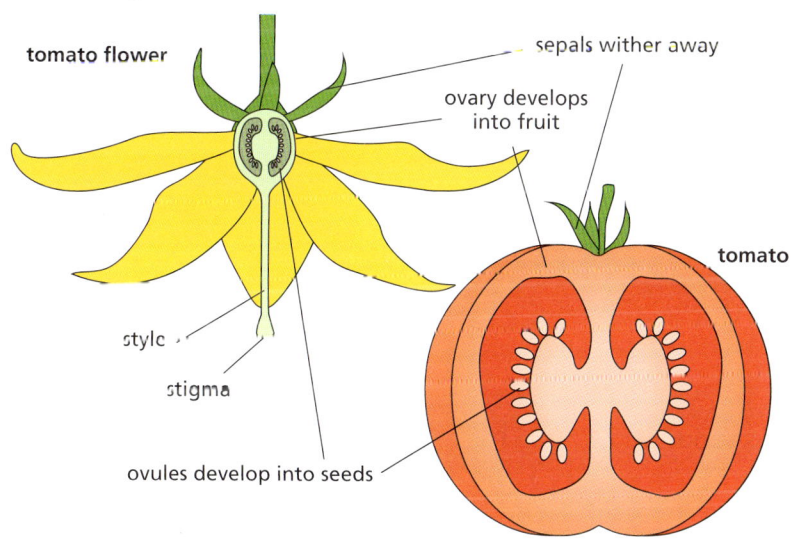

FIGURE 1.9.6c: How a flower develops into a fruit.

5. Make a table and record what happens to the petals, stamen, ovule and ovary following fertilisation.
6. Suggest why pea pods are fruits although they are often called and classed as a vegetable.

Know this vocabulary

fertilisation
pollen tube
seed
fruit

SEARCH: pollination and fertilisation in flowering plants

Ecosystems

Understanding how seeds are dispersed

We are learning how to:
- Recognise the variety of different structures of different seeds.
- Describe the need for plants to disperse their seed.
- Plan an investigation into seed dispersal by wind.

The largest seed in the world is 50 cm in diameter. It comes from the palm tree called Coco de Mer, found only in the Seychelles islands in the Indian Ocean. Another large seed is the coconut – it can be carried by the sea and germinate in a new place. Plants have developed many ingenious ways to be dispersed and to colonise new areas.

The challenge of moving seeds

Plants colonise new areas by moving their seeds in a process called **dispersal**. Seeds can be dispersed by:
- wind;
- water;
- exploding pods that release seeds on touch or with moisture;
- being carried inside animals that eat the fruit;
- hooking onto the fur or skin of passing animals.

1. Look at Figure 1.9.7b, and identify how each of the three seeds is dispersed.
2. Give reasons for all of your answers to question 1.

FIGURE 1.9.7a: Coco de Mer seed – the largest on the planet.

Ways of travelling

Seeds dispersed by wind have many shapes and sizes. The dandelion has parachute-like seeds, and the sycamore has seeds like helicopters. Peas and pansies have pods that explode when they have dried out or are touched by an animal, causing the seeds to fly out. Some plants produce fruits that animals eat but cannot digest. These pass through the animals, allowing the seed to begin **germination** in another place, using nutrients from the animals' dung. Burdock seeds have tiny hooks that catch on the fur of passing animals.

FIGURE 1.9.7b: Why have plants developed such variety in types of seed?

Seeds need to be dispersed as far away from the parent plant as possible, where there could be more light, nutrients and water – thereby increasing the chance of successful growth. Seeds are packed with nutrients to help the germinating plant to grow. Smaller seeds have fewer nutrients but may travel further. Larger seeds have bigger stores of food and can last much longer.

9.7

FIGURE 1.9.7c: How are these seeds dispersed?

3. Why are the seeds from trees in forests most likely to be dispersed by the wind?
4. What are the advantages and disadvantages of a seed growing near the parent plant?
5. Suggest why the coconut seed is carried by water rather than by air.

Did you know…?

The seeds of the *Alsomitra* vine tree were the inspiration in the development of the first gliders and aeroplanes. With a wing span of up to 13 cm, they are the largest wind-pollinated seeds in the world.

Investigating models of seed dispersal

The sycamore seeds shown in Figure 1.9.7c are often described as travelling like mini helicopters. Their small size and aerodynamic features allow them to be dispersed over large areas. We can model this seed and investigate what affects its dispersal.

Make a model seed from paper, like that in Figure 1.9.7d. Consider:
- How can you measure how well it can be dispersed?
- What factors might affect how well it is dispersed?

6. Plan an investigation into a factor that affects how well this model seed works. Consider the **independent**, **dependent** and **control variables**. Plan what you will measure and how you will record your results.
7. Draw a conclusion based on your investigation. If possible, compare your model to real sycamore seeds.
8. Describe the most successful seed, based on your modelling. Could you design other seeds that would be successfully dispersed?

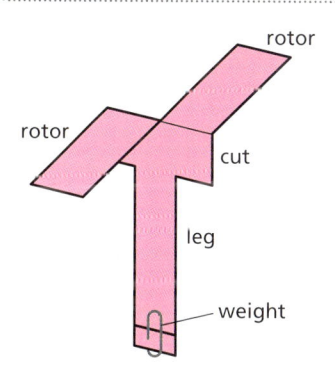

FIGURE 1.9.7d: A model of a paper mini helicopter.

Know this vocabulary

dispersal
germination
independent variable
dependent variable
control variable

SEARCH: wind dispersal of seeds 213

Ecosystems

Understanding how fruits disperse seeds

We are learning how to:
- Describe how fruits are used in seed dispersal.
- Compare evidence about seed dispersal by wind and fruit formation.
- Use data to evaluate different seed dispersal mechanisms.

Without animals to disperse their seeds, some plants would become extinct. The seeds of the *Astrocaryum* palm used to be dispersed by dinosaurs. Now, small rodents called agoutis disperse the seeds. Agoutis steal each other's seeds, increasing the distance of dispersal.

Plants exploiting animals

Plants may develop edible fruits to disperse seeds. A **fruit** is the ovary of a plant after fertilisation. The fruit is a nutritious treat surrounding the seed or seeds, mainly made of sugars and tasty nutrients to attract animals. Examples include some nuts, tomatoes and cucumbers. Some seeds cannot be digested, so pass through the intestines and out with the faeces. Some seeds, such as mango seeds, are too large to be eaten. When they are discarded on soil, they can germinate to make new plants.

Fruit contains lots of energy, which is transferred to the animals that eat the fruit. Plants that disperse seeds by fruit do not need to produce as many seeds, as most are carried away from the parent plant and end up in soil that is nutrient-rich from dung.

1. What is a 'fruit'?
2. Which of the items in Figure 1.9.8a is not a fruit?
3. What is the main advantage of fruits dispersing seeds?

FIGURE 1.9.8a: Where are the seeds in these plant products?

Surveying and sampling seeds

Botanists carry out surveys to try to find out how seeds are dispersed and how successful different plants are at germinating the seeds they make. They might do this by sampling many plants of the same species in a particular habitat. First they count the number of seeds made. Then, after the seeds have dispersed, they sample the habitat again to make an estimate of the number of seedlings that have germinated. By comparing the number of seeds that have germinated with the number of seeds made originally, they can judge how successful the seed dispersal mechanism is.

4. What is the independent variable in this survey?
5. Which variables need to be controlled in such a survey?
6. How would you ensure the evidence collected was **reliable**?
7. Why might it be important to find out how successful plants are at dispersing and germinating seeds?

How efficient are different methods of seed dispersal?

TABLE 1.9.8: Different methods of seed dispersal.

Name of plant	Type of dispersal mechanism	Approximate number of seeds made per plant	Average dispersal distance
ragwort	parachute	10 000	over 100 m
ash tree	helicopter	1000	over 100 m
Alsomitra vine tree	glider	40 000	1–2 km
witch hazel	exploding pod	100	10 m
pea	exploding pod	100	a few metres
blackcurrant	fruit	300	variable
melon	fruit	500	variable
coconut	water	50	hundreds of miles

Table 1.9.8 summarises the types of seed dispersal mechanisms used by a variety of plants.

8. If you were a plant, which dispersal mechanism would you prefer and why?
9. What can you say about the different dispersal mechanisms from the data?
10. Show the data from the table in a graphical form. Choose a good way to represent the data so that the different mechanisms can be evaluated.

Did you know...?

Avocados are thought to be the most nutritious fruit, with over 25 essential nutrients, including vitamin C, iron, magnesium and potassium. The demand for avocados has led to the destruction of some of Mexico's pine forests as farmers make space to grow as many avocados as possible.

Know this vocabulary

fruit

SEARCH: seed dispersal by fruits

Ecosystems

Checking your progress

To make good progress in understanding science you need to focus on these ideas and skills.

- Describe an example of a simple food web.
- Define producers, consumers and decomposers and give examples of each in different food webs.
- Describe how changes in the population of one organism can influence other organisms in the food web.

- Describe the role of insects in fruit crop production.
- Explain why artificial pollination is used for some crops.
- Explain what is meant by 'food security' and explain the risks posed by monoculture on food security.

- Recall ways in which organisms can affect their environment.
- Explain how changes in predator and prey populations affect each other.
- Use data and models to predict changes to predator and prey populations based on their interdependence.

- Give examples of toxins and describe how toxins pass along a food chain.
- Explain how toxins accumulate in food chains.
- Evaluate the advantages and disadvantages of using pesticides

9.9

- [ ] Describe the roles of different parts of a flowering plant in reproduction.

- [ ] Explain the differences in wind-pollinated and insect-pollinated plants.

- [ ] Discuss the strengths and weaknesses of wind pollination and insect pollination.

- [ ] Recognise that pollination and fertilisation are both part of plant reproduction but are two different processes.

- [ ] Describe the stages of fertilisation in plants, including the role of the pollen tube.

- [ ] Describe the fate of flower structures following fertilisation and the formation of seeds and fruit.

- [ ] Recognise different seed-dispersal methods and relate these to the structures of the seeds.

- [ ] Identify key variables that need to be controlled when investigating the effect of seed design on seed dispersal.

- [ ] Explain the advantages and disadvantages of different seed-dispersal mechanisms.

Ecosystems

Questions

KNOW. Questions 1–6

See how well you have understood the ideas in this chapter.

1. What is the name for an animal that eats other animals or plants? [1]
   a) Producer   b) Decomposer   c) Consumer   d) Prey

2. Why does preying on more than one animal help populations to survive? [1]
   a) Toxins are less likely to get into the food chain.
   b) It allows animals to have a more varied diet.
   c) If the population of one prey decreases, there is still a food source for the predator.
   d) It allows the predator to move higher up the food chain.

3. Describe why bee numbers have decreased and explain the effect of reduced numbers of bees on crop production. [3]

4. Which structure is not directly linked to fertilisation? [1]
   a) Ovule   b) Ovary   c) Stigma   d) Pollen grain

5. The male parts of a flowering plant as a whole are called: [1]
   a) Carpel   b) Stigma   c) Stamen   d) Pollen

6. Describe the events that take place after pollination, leading to a fruit being formed. [4]

APPLY. Questions 7–12

See how well you can apply the ideas in this chapter to new situations.

7. Harmful algal blooms (HABs) produce toxins in the sea. Oysters are animals that filter food particles like plankton from the water. How could dining on oysters during a HAB affect a person's health? [1]

8. Look at this simple food web in a rainforest. What will happen to the number of red-eyed tree frogs if all the chimpanzees die from a disease? [1]
   a) They stay the same.
   b) They go up.
   c) They go down.
   d) They will die out too.

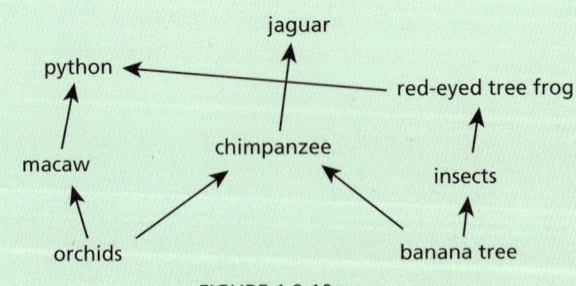

FIGURE 1.9.10a

9. Look at the food web in Figure 1.9.10a. The jaguar and the python are predators. What will happen if both of these predators die out? [4]

218 AQA KS3 Science Student Book Part 1: Ecosystems – Interdependence *and* Plant reproduction

10. Some plants live in areas of high density of other plants. Choose **two** ways that plants may compete with others to attract pollinating insects. [2]

   a) They have brightly coloured flowers.
   b) They have high levels of chlorophyll in their leaves.
   c) They produce huge numbers of light, feathery pollen grains.
   d) They produce large amounts of nectar.

11. Seeds are dispersed by a variety of mechanisms. Some are shown in Figure 1.9.10b. Which type of seed is likely to be dispersed by: [2]

   a) water?
   b) being carried on the fur of an animal?

burdock seed   avocado stone   dandelion head   coconut

FIGURE 1.9.10b

12. A rare fruit and its seeds are analysed and are found to contain large amounts of energy compared to several other fruits. Suggest **two** reasons why containing lots of energy supports the growth of a new fruit plant. [2]

EXTEND. Questions 13–14

See how well you can understand and explain new ideas and evidence.

13. Figure 1.9.10c shows how the populations of lynx and hares changed over time. Analyse and evaluate the data to explain why the populations rise and fall when they do. Do you think this pattern is still happening today? Explain your answer. [4]

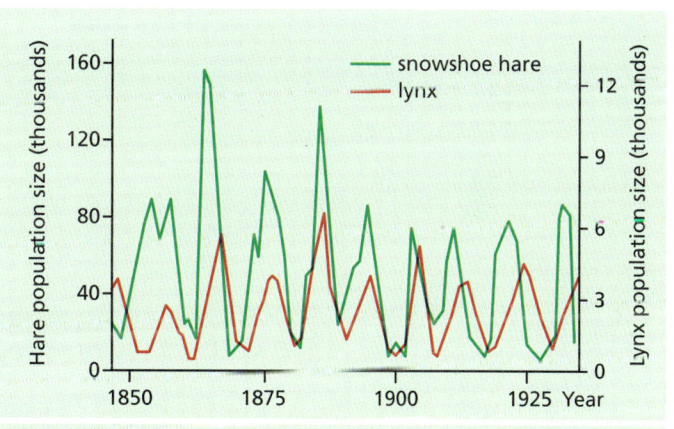

FIGURE 1.9.10c: Lynx and hare population data.

14. Using the average data in Table 1.9.10, describe the effect of sugar concentration on growth of pollen tubes. Suggest which one result the students ignored when calculating the averages. [3]

TABLE 1.9.10: The growth of pollen tubes in different sugar concentrations.

Sugar concentration (%)	5	10	15	20
Growth of pollen tubes (micrometres) – experiment 1	225	345	200	213
Growth of pollen tubes (micrometres) – experiment 2	250	350	450	207
Growth of pollen tubes (micrometres) – experiment 3	275	355	450	250
Average growth of pollen tubes	250	350	300	233

Genes
Variation *and* Human reproduction

Ideas you have met before

Variation and classification
Living things are classified into broad groups according to observable characteristics, similarities and differences.

Adaptations
Animals and plants are adapted to the conditions of the habitats in which they live.

An adaptation is a way an animal's body helps it survive in its environment – for example meerkats have dark circles around their eyes, which act like sunglasses, helping them see even when the Sun is shining very brightly.

Human reproduction and development
The gametes in animals are the egg cell and the sperm cell.

Fertilisation happens when the nucleus of a male gamete fuses with the nucleus of a female gamete.

Humans change throughout their lifetime, from the moment of conception to the time they grow old.

Some changes occur much faster than others. We change fastest during the first few months of our existence.

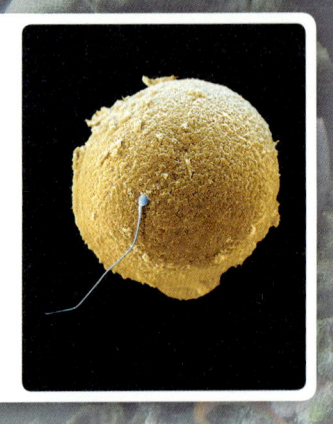

10.0

In this chapter you will find out

Variation

- There is variation within a species and this can be measured and classified as continuous or discontinuous variation.
- Variations can be caused by the environment or by inheritance, but many are caused by a combination of both factors.
- Variation between organisms ensures that some organisms survive.
- Species that have too little variation may become extinct.

Human reproduction

- The male and female human reproductive systems are adapted for successful reproduction.
- Puberty and reproduction are controlled by hormones. Drugs can be used to support infertility and contraception.
- When an egg is fertilised it develops into a foetus. This grows in the uterus until it becomes a fully grown baby.
- Many factors affect the growth and development of a foetus, including the mother's use of alcohol, cigarettes and drugs.

Genes

Looking at variation

We are learning how to:
- Describe what is meant by variation in a species.
- Explain the difference between continuous and discontinuous variation.
- Plot graphs to show variation.

Look around you. What differences can you see between the people in your class? **Variation** in characteristics can be classified in different ways and graphs can help us to understand the different patterns of variation.

FIGURE 1.10.1a: What variation can you see among these children?

Spot the difference

A **species** is a group of living things that have more in common with each other than with other groups of living things. Organisms within a species are able to reproduce and produce fertile offspring. The scientific name for the human species is *Homo sapiens*, the dog species is called *Canis familiaris*. Within species, there are many features that vary; some of these are obvious to see, such as height, hair colour or leaf shape. Other features are not easy to see, such as blood group and differences in our genes.

1. Define 'species'.
2. List five things that differ among humans and five things that differ among dogs.

a labradoodle puppy

FIGURE 1.10.1b: A labradoodle puppy with its parents – a labrador retriever and a standard poodle.

Types of variation

The height of a human population ranges from the shortest to the tallest individuals (called the range) – any height is possible between these values. A feature that changes gradually over a range of values is said to have **continuous variation**. Examples of such features are height and wing span.

The bell-shaped graph showing height values (Figure 1.10.1c) is called a 'normal distribution'. This is what you would expect to find in any feature with continuous variation. The most frequently occurring value is called the mode.

Some features have only a limited number of values. An individual has one type of the feature or another. This is called **discontinuous variation** – examples of this are gender, blood group (Figure 1.10.1d) and vein patterning in leaves. Measuring and recording variation can help us to understand which type of variation a feature shows.

Did you know…?

Lions and tigers can be cross-bred to produce ligers and tigons. These offspring are usually infertile (cannot breed) and so lions and tigers are classed as separate species. Although beautiful, both ligers and tigons often have serious health problems.

3. For each feature you have identified in question 2, state whether it is an example of continuous variation or discontinuous variation.
4. Explain the difference between continuous and discontinuous variation.
5. Look at the data collected in Table 1.10.1a and Table 1.10.1b on tongue rolling and arm span. Plot a graph for each of these data. Which type of variation does each feature show?

TABLE 1.10.1a

Tongue rolling	Number of students
Can roll tongue	19
Cannot roll tongue	11

TABLE 1.10.1b

Arm span (cm)	Number of people
146–150	3
151–155	16
156–160	20
161–165	24
166–170	22
171–175	15
176–180	8

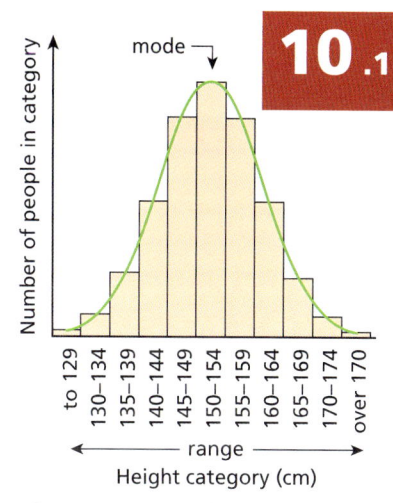

FIGURE 1.10.1c: Height shows normal distribution.

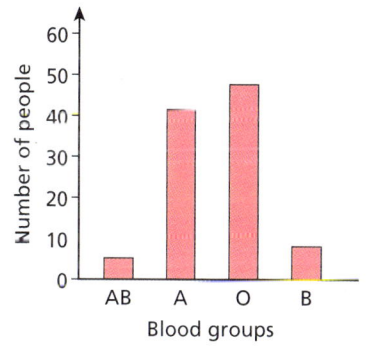

FIGURE 1.10.1d: Blood groups show discontinuous variation.

Investigating variation

Scientists can investigate variation to find out if features are linked, such as students' height and shoe size. The larger the sample size used, the more reliable the data.

A scattergraph is used to show whether or not there is a relationship between two sets of data. The graph may show:

- a positive **correlation** – one quantity increases as the other does (as in Figure 1.10.1e);
- a negative correlation – one quantity increases as the other decreases;
- no correlation – there is no clear relationship.

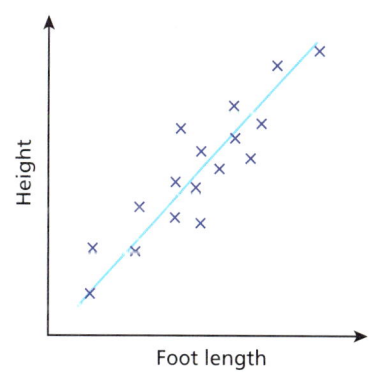

FIGURE 1.10.1e: There is a positive correlation between height and foot length in humans.

6. Describe an investigation to see if there is a link between the length of a holly leaf and the number of spikes on the leaf.
7. Explain why sample size is important.

Know this vocabulary

variation
species
continuous variation
discontinuous variation
correlation

SEARCH: continuous and discontinuous variation

Genes

Exploring causes of variation

We are learning how to:
- Identify whether variation is caused by inheritance or by environmental factors.
- Understand that offspring from the same parents may show variation.

There are millions of plants and animals on Earth. They are all different from one another. What causes these differences? Why are some organisms almost identical?

What causes variation?

One cause of variation in organisms is their environment. For example your diet, health and the amount of exercise you do affect your growth. Climate and food supply influence all living things. When animals fight (for example over available resources), they may lose teeth or develop scars from deep wounds. A person's country and culture can also be sources of variation – for example Buddhist monks shave their heads.

FIGURE 1.10.2a: The foal inherits the length of its legs from its parents.

The other major cause of variation in organisms is the passing on of features from parents to offspring, and from their parents before them, and so on. **Inherited** variations are **genetic** and cannot be altered. For example, you may dye your hair purple, but it will always grow back in your natural colour. Sometimes there are clear traits that run in families, such as the shape of your nose or the presence of a dimple or freckles.

1. Name one characteristic that you have inherited from your parents, and one caused by the environment.
2. Name three features that a foal inherits from its parents.

Did you know…?

Even twins can show huge variation. The girls in the picture are twins. Due to inheriting a mix of genes from both their mother and father, they were born with different hair colouring.

Why are offspring different?

Brothers and sisters from the same parents can look very different from each other. Parents pass on genetic information to their offspring in the nucleus of their sex cells. The offspring inherit one set of information from the mother's egg cell nucleus and one set from the father's sperm cell nucleus. Every egg cell and every sperm

FIGURE 1.10.2b

224 AQA KS3 Science Student Book Part 1: Genes – Variation *and* Human reproduction

cell contains different hereditary information from the respective parent. Each fertilised egg cell will therefore contain a different, random and unique combination of characteristics inherited from the parents. Offspring may have some similarities to their siblings, but may also look very different.

3. Why can children from the same parents be very different from each other?

4. Look carefully at Figure 1.10.2c.
   a) How are the children similar to each other?
   b) How are they different from each other?

FIGURE 1.10.2c. What variation can you see in this family?

Genetic or environmental?

Variation in characteristics is needed for the process of evolutionary change that enables a species to change gradually and survive. Sometimes this results in the development of a new species.

Scientists have debated whether certain characteristics are inherited or are caused by the environment in which people live. They now generally agree that only a small number of human features are entirely inherited and are not in any way affected by the environment. These features include:

- natural eye colour;
- natural hair colour;
- blood group;
- some inherited diseases.

Some features have a well-established genetic basis; others are mostly due to the environment. However, most features are caused by the interaction of genetic and environmental factors. For example, a person's skin may have birthmarks and moles, but during their lifetime scars may form and tattoos may be added.

FIGURE 1.10.2d: Tattoos are important in many cultures.

5. Give three other examples of inherited human features that can be affected by the environment in which a person lives.

6. Evaluate the importance of genetic and environmental variation in the survival of an organism.

Know this vocabulary
inherited
genetic

SEARCH: causes of variation

Genes

Considering the importance of variation

We are learning how to:
- Describe the importance of variation.
- Explain how variation may help a species to survive.
- Apply ideas about variation and survival to specific examples.

Differences between living things can help a species to survive. Why is variation so important to survive a changing environment?

The importance of variation

Variation is important for the survival of a species. If all organisms were identical and the environment changed so that none of the organisms were adapted to survive, they would die out. For example, the dodo bird lived on the island of Mauritius until 1681. Without any predators throughout its history, the dodo became a flightless bird, with no variation in flying ability. The introduction of humans and other mammals to the island led to the hunting of the dodo and, with no way of escaping predators, the dodo population decreased until it became **extinct**. If there had been variation in how well dodos could fly, there could have been an opportunity for the species to survive.

FIGURE 1.10.3a: A dodo.

1. Explain why the dodo species died out.

FIGURE 1.10.3b: What might happen to the shorter giraffe?

Surviving change

Variation in a population gives those organisms with more favourable features a **survival advantage**. When these organisms reproduce, the feature is passed on to offspring.

Some variation benefits a species, for example:

- Not all rabbits are killed by the viral disease myxomatosis. Some rabbits are resistant to infection and can survive an outbreak.
- Peacocks with the best display of feathers are most likely to attract a mate.

Some differences are not beneficial, for example:

- Albino (pure white) giraffes do not survive long in the wild.
- Antelopes that run slower than the herd do not survive.

FIGURE 1.10.3c: Some animals are born with no pigment. They are called albinos.

10.3

2. Look at Figures 1.10.3b and 1.10.3c. Describe the variation in the giraffes and in the squirrels and suggest any survival advantage in each case.

3. Look at the moths in Figure 1.10.3d. In unpolluted conditions, the tree bark on which these moths sit are pale with lichen. In polluted conditions, the tree bark becomes dark. Describe how variation helps these moths to survive as the amount of pollution changes.

Superbugs »»»

Variation is also seen in micro-organisms, such as bacteria. Some bacteria are resistant to some antibiotics. When antibiotic medicines are used, any bacteria that are resistant have the advantage and are more likely to survive and reproduce to form more resistant bacteria. As the use of antibiotics has increased, so too has the number of bacteria that are resistant to the drugs.

Antibiotic resistance is a huge problem in hospitals. There are some strains of bacteria that are resistant to many different antibiotics – these are called superbugs. One superbug is MRSA (methicillin-resistant *Staphylococcus aureus*). An infection by MRSA can be very difficult to treat.

FIGURE 1.10.3d: A dark and pale Peppered Moth (*Biston betularia*).

4. Explain how variation in antibiotic resistance helps species of bacteria to survive.

5. Look at Figure 1.10.3e. Suggest why prescribing methicillin has become less effective between 1982 and 2014.

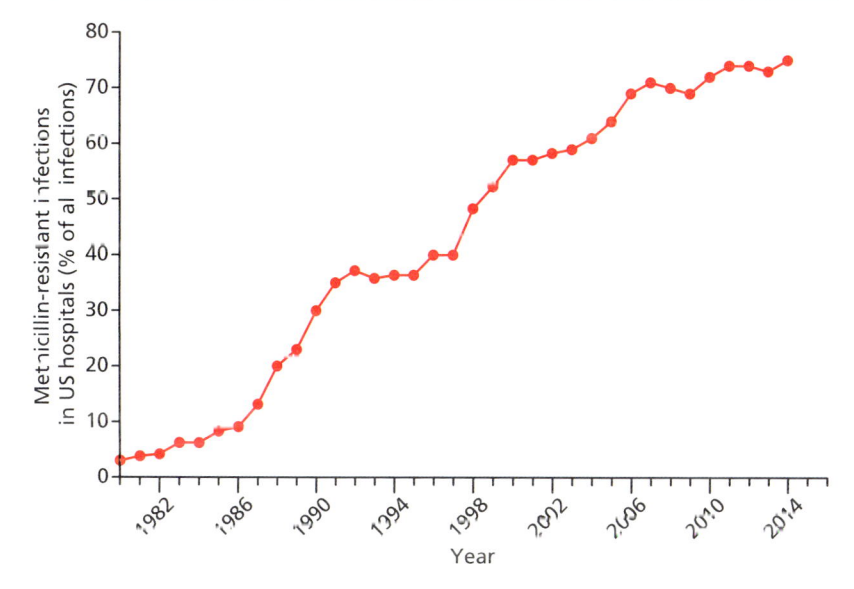

FIGURE 1.10.3e: Antibiotic resistance in bacteria.

Did you know…?

Following the discovery of the superbug MRSA, hospital doctors were required to stop wearing ties and long-sleeved shirts. It was thought that their clothing was carrying bacteria from one patient to another.

Know this vocabulary

extinct
survival advantage

SEARCH: examples of survival advantage   227

Genes

Understanding the female reproductive system and fertility

We are learning how to:
- Describe the structure and function of different parts of the female reproductive system.
- Describe the process of menstruation.
- Describe causes of low fertility.

The human female fertility cycle is controlled by chemicals called hormones. The female reproductive system receives sperm and enables the fertilised egg to develop until it is ready to be born. The uterus, or womb, is where the foetus grows and develops. The uterus increases to up to 20 times its original size during pregnancy.

The functions of female organs

The human female **reproductive system** has two main purposes – to produce egg cells that may be fertilised by the male sperm, and to provide an environment for the growing foetus.

The main female organs are the **vagina**, **cervix**, **uterus**, **oviduct** and **ovary**. Table 1.10.4 summarises the structure and function of each of these.

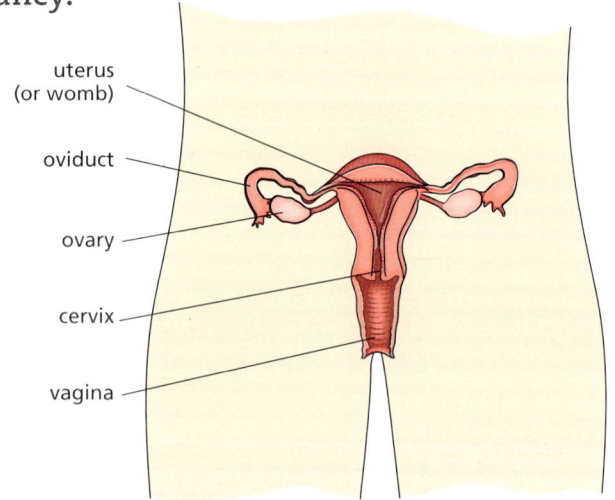

FIGURE 1.10.4a: The female reproductive system.

TABLE 1.10.4: Female reproductive organs.

Vagina	Muscular tube, 8 to 12 cm long, that extends up to the uterus and can stretch to allow a baby to pass.
Cervix	Narrow opening from the vagina to the uterus with thick walls – can extend wide enough to allow a baby to pass.
Uterus or womb	Pear-shaped cavity with thick muscular walls, where the developing baby grows.
Oviduct (Fallopian tube)	The tube that carries the egg from the ovary to the uterus.
Ovary	Where eggs cells are made and then released into the oviduct.

1. Where are female sex cells made?
2. Why do you think the uterus has muscular walls?

10.4

Menstruation

Menstruation occurs in a cycle lasting about 28 days and is controlled by hormones.

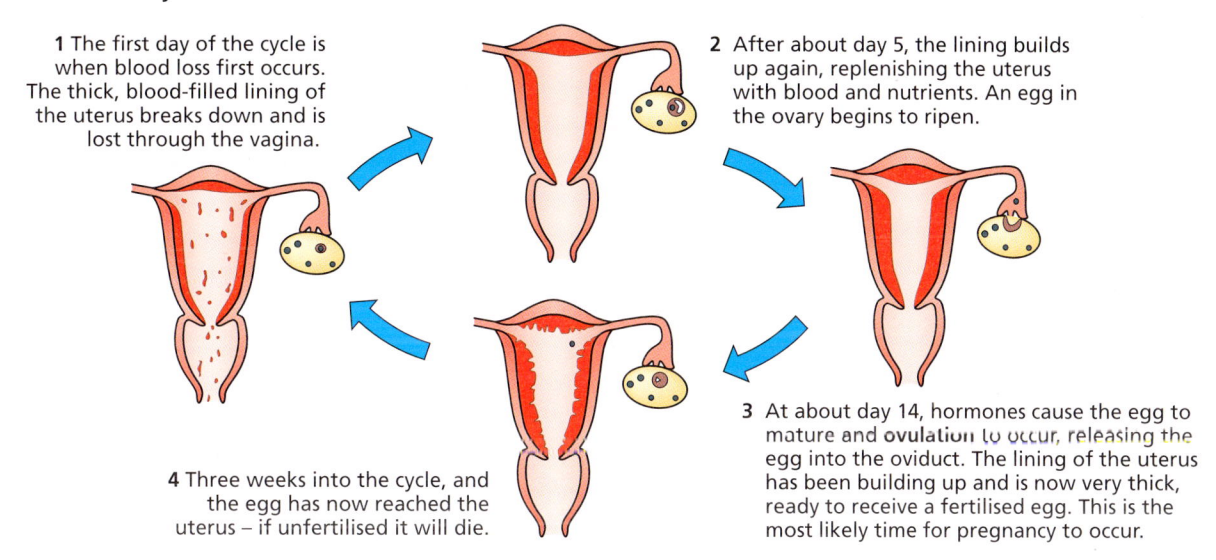

1. The first day of the cycle is when blood loss first occurs. The thick, blood-filled lining of the uterus breaks down and is lost through the vagina.

2. After about day 5, the lining builds up again, replenishing the uterus with blood and nutrients. An egg in the ovary begins to ripen.

3. At about day 14, hormones cause the egg to mature and **ovulation** to occur, releasing the egg into the oviduct. The lining of the uterus has been building up and is now very thick, ready to receive a fertilised egg. This is the most likely time for pregnancy to occur.

4. Three weeks into the cycle, and the egg has now reached the uterus – if unfertilised it will die.

FIGURE 1.10.4b: The menstrual cycle.

3. What part do hormones play in menstruation?

Infertility

Most women below the age of 36 have little trouble in having babies. However, **infertility** affects about 3.5 million women in the UK. There are a number of causes:

- External factors – such as excessive alcohol, drugs, long-term smoking, stress and sexually transmitted diseases.
- Problems with **ovulation** – the release of eggs is controlled by hormones; a hormonal imbalance may result in eggs not being made or released.
- Endometriosis – cells from the lining of the oviduct may start to grow around the ovary and cause cysts to appear, making it difficult for the eggs to be released.
- Blockages in the oviduct – these can prevent an egg from reaching the uterus and becoming fertilised.

In men, infertility may be caused by a low number of healthy sperm or sperm that can't swim well because of disease.

Some fertility drugs work by increasing the number of eggs that a woman produces and releases each month. This increases the chance of pregnancy. Contraceptive drugs can work by stopping ovulation and so, with no egg, the pregnancy cannot occur.

4. For each of the female infertility problems, suggest a possible solution.

Did you know…?

The ovaries of new-born girls have about 600 000 immature eggs. However, an adult woman is capable of giving birth to a maximum of 35 babies.

Know this vocabulary

reproductive system
vagina
uterus
oviduct
ovary
menstruation
infertility
ovulation

SEARCH: human female reproductive system

Genes

Understanding the male reproductive system and fertilisation

We are learning how to:
- Describe the structure and function of different parts of the male reproductive system.
- Describe fertilisation in humans.

The human reproductive systems are controlled by chemicals called hormones. In the male, one hormone is testosterone, which controls the growth and development of the organs and sperm cells. Sperm cells take four to six weeks to mature, and live for about 36 hours once released inside the female.

The functions of male organs

The purpose of the human male reproductive system is to produce millions of male sex cells (sperm) and to transport them inside the female to fertilise an egg cell and so produce a baby.

The main parts of the male reproductive system are the **testicles**, scrotal sac, **sperm duct**, prostate gland, **semen**, **urethra** and **penis**. Table 1.10.5 summarises the structure and function of each of these.

TABLE 1.10.5: Male reproductive organs.

Testicles	Two organs where sperm cells are made.
Scrotal sac	Protection around the testicles. This sac holds the testicles outside the body where sperm is kept at the best temperature for them to function.
Sperm duct	The tube that carries the sperm from the testicles to the prostate gland.
Prostate gland	Where semen is made.
Semen	A liquid that mixes with sperm and provides them with nutrients for their journey.
Urethra	The tube leading from the prostate gland along the penis.
Penis	The organ around the urethra. Movement of the penis releases sperm during intercourse.

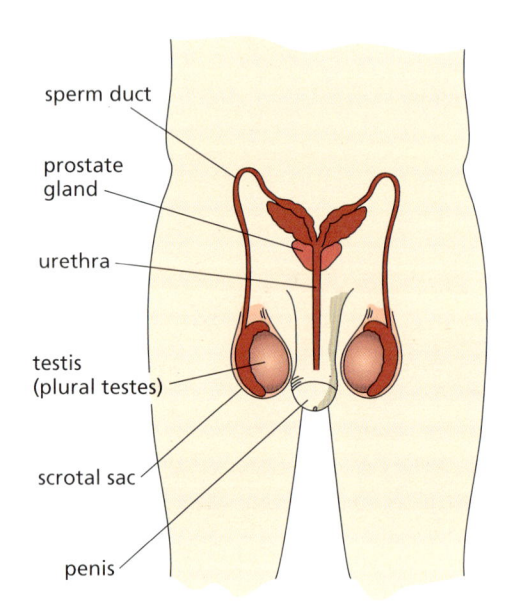

FIGURE 1.10.5a: The male reproductive system.

Did you know...?

A human sperm is the smallest cell in the body. 5000 sperm cells would fit into one millimetre. The egg cell is the largest – about the size of a full stop.

10.5

1. List one cell and two organs in the male reproductive system.
2. Draw the journey of a sperm cell, labelling the parts of the male reproductive system that it passes through.

Transfer of the male sex cell

Humans carry out internal fertilisation. In sexual intercourse, the penis is inserted inside the vagina, and its movement stimulates the release of sperm from the testicles. In this way, sperm are guaranteed to be placed directly inside the female. Both a plant's anthers and the testicles produce millions of male sex cells to maximise the likelihood of successful fertilisation. However, plants produce pollen only when the stigmas are likely to be ready for fertilisation.

FIGURE 1.10.5b: The human male sex cell is adapted to carry out its job. How is it different from a pollen cell?

3. Which parts make the male sex cells in plants and in humans?

Fertilisation

The general name for the sex cells (egg and sperm) is **gametes**. One egg matures each month in the ovary and is released into the oviduct – this process is ovulation. The lining of the oviduct contains specialised cells with tiny hairs that beat, causing the egg to move down to the uterus. For up to 24 hours the egg may be fertilised by a sperm cell. Only one sperm penetrates the egg cell, losing its tail as it does so. **Fertilisation** is when the nucleus of the sperm fuses with the nucleus of the egg, combining the genetic material of both. The fertilised egg is the start of a new life. Once the egg is fertilised, the menstrual cycle stops to allow the foetus to develop.

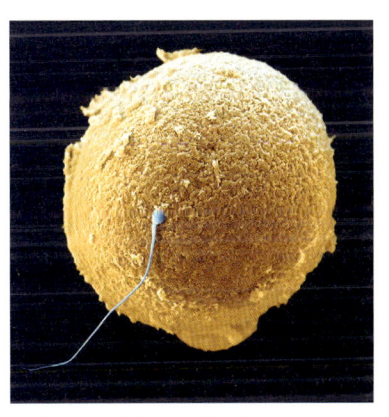

FIGURE 1.10.5c: Fertilisation occurs when one sperm cell penetrates the egg cell and their nuclei fuse.

Know this vocabulary

testicle
sperm duct
semen
urethra
penis
gamete
fertilisation

4. Why does the egg need to move from the ovary to the uterus?
5. Why do you think the egg cell is so much bigger than the sperm cell?
6. Explain the difference between ovulation and fertilisation.

SEARCH: human male reproductive system

Genes

Learning how a foetus develops

We are learning how to:
- Describe the role of the mother in supporting and protecting the developing foetus.
- Describe the stages in the development of a foetus.

A human foetus takes 38 weeks to grow from one fertilised cell into a complete baby ready to be born. Dogs take just two months, whereas elephants take up to two years. This period of development is called **gestation**. The mother provides the developing foetus with all the nutrients and oxygen it needs, as well as removing all waste products.

Cell division

When an egg cell has been fertilised, it divides into two cells. These cells further divide to make four cells, which divide again to make eight cells. This cell division continues until there are several thousand cells. This is the process of growth, where cells divide to make new cells and the overall size of the organism increases. Within the first two to three weeks the cells are all the same – they are called stem cells, and have the ability to become any specialised cell in the body.

1. What is 'growth'?
2. What is special about stem cells?

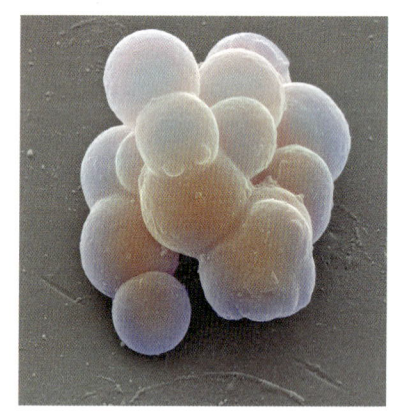

FIGURE 1.10.6a: Stem cells.

Development of the foetus

Once the ball of stem cells reaches a certain size, the cells begin to differentiate and become specialised cells. Some cells will develop into the organs and tissues of the developing baby. At this stage when the cells begin to differentiate, the ball of cells is called an **embryo**. Once it reaches about 8 weeks old, when most of the main organs are formed, including the heart which is now beating, it is called a **foetus**.

Figure 1.10.6b shows the different stages of development of a human foetus. Ultrasound is used to make images of the foetus at different stages to monitor its development and identify any problems. The size of the foetus can be measured using these images.

3. When is the fastest period of growth of the developing foetus? Explain your answer.

Did you know…?

The taste buds of a foetus develop at 14 weeks; it can hear at 24 weeks and track objects with its eyes at 31 weeks. At 28 weeks, a foetus is likely to survive if born.

10.6

weeks of gestation	8	12	16	20	24	28	32	36	40
size (length) of foetus	40 mm	100 mm	140 mm	190 mm	230 mm	270 mm	300 mm	340 mm	380 mm

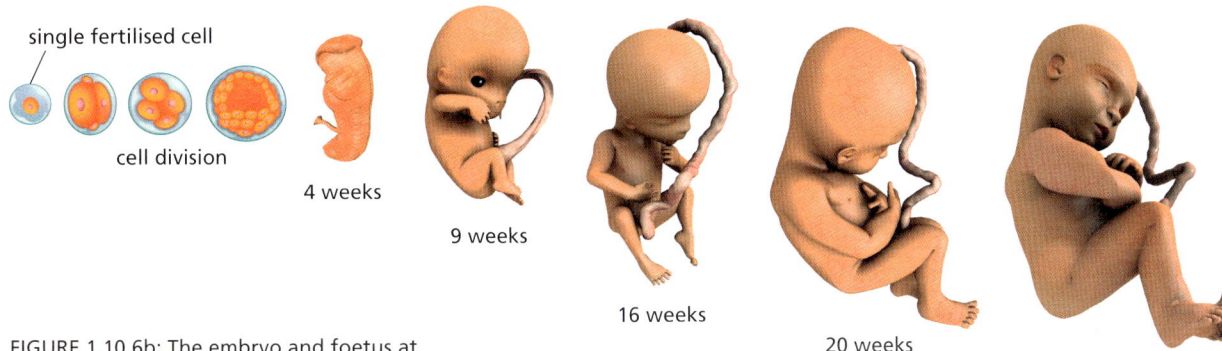

FIGURE 1.10.6b: The embryo and foetus at different stages of development.

Supporting structures

During pregnancy, other cells from the original ball of cells will become structures that connect with the mother – the **placenta**, amnion, **amniotic fluid** and **umbilical cord**. These structures are shown in Figure 1.10.6c.

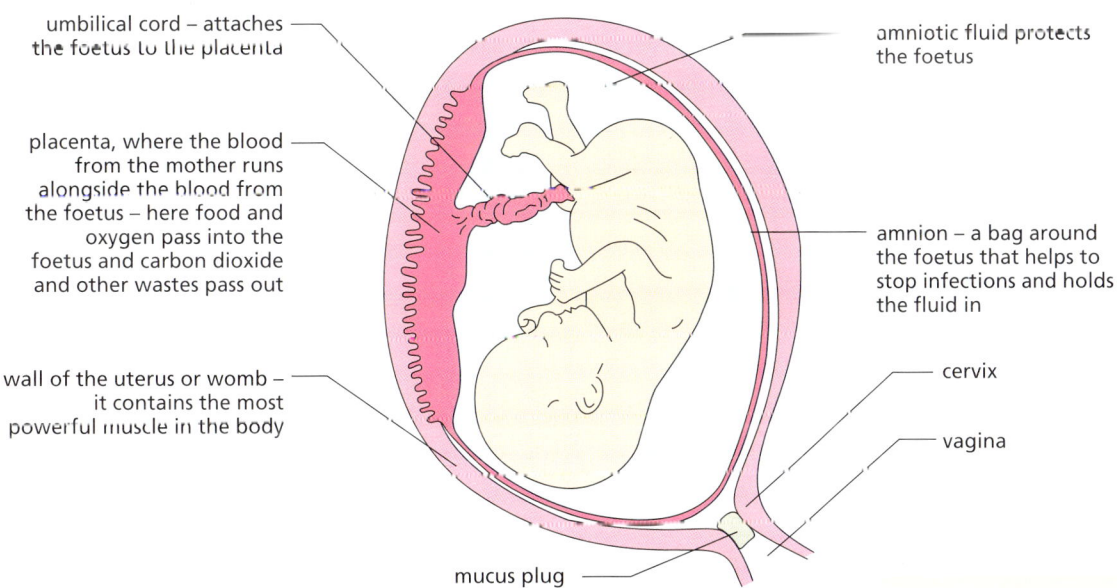

- umbilical cord – attaches the foetus to the placenta
- placenta, where the blood from the mother runs alongside the blood from the foetus – here food and oxygen pass into the foetus and carbon dioxide and other wastes pass out
- wall of the uterus or womb – it contains the most powerful muscle in the body
- amniotic fluid protects the foetus
- amnion – a bag around the foetus that helps to stop infections and holds the fluid in
- cervix
- vagina
- mucus plug

FIGURE 1.10.6c: The developing foetus in the uterus.

4. Why does a foetus need the placenta?
5. Why is it important for the baby to be surrounded by fluid?
6. Summarise the different ways in which a pregnant uterus is different from a normal uterus.

Know this vocabulary

gestation
embryo
foetus
placenta
amniotic fluid
umbilical cord

SEARCH: development of a human foetus

Genes

Understanding factors affecting a developing foetus

We are learning how to:
- Describe the effects of different factors on a developing foetus.
- Evaluate the strength of data.
- Analyse advice given to pregnant women.

A foetus cannot take in its own food or oxygen and relies on the mother to supply it with essential chemicals and nutrients. The placenta allows substances to pass from mother to baby.

The role of the placenta

The placenta allows oxygen, glucose, digested proteins and fats, vitamins and minerals to enter the foetus – it also removes carbon dioxide and waste products, such as urea. Harmful substances can also cross the placenta including alcohol, nicotine, carbon monoxide, cocaine, insecticides, lead and mercury.

1. How might the harmful substances come to be at the placenta?
2. What would happen to a foetus without the placenta? Explain your answer.

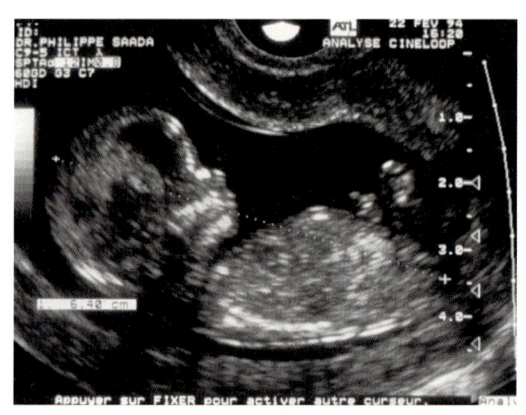

FIGURE 1.10.7a: An ultrasound scan of a foetus enables its development to be checked.

Effects of substances on the foetus

Scientific studies have established how different substances affect a developing foetus. Foetal size and movements can be tracked and the heartbeat measured. Tests have found out how some substances affect the foetus – see Table 1.10.7.

TABLE 1.10.7: Substances that affect a developing foetus.

Alcohol	Higher rate of stillbirth (baby dead at birth), lower birth weight, lower IQ; baby slower to move and think; more likely to be dependent on alcohol in adulthood.
Smoking – nicotine and carbon monoxide	Much higher risk of stillbirth, **premature** birth and low birth weight resulting in poor development; greater likelihood of developing asthma.
Drugs – marijuana, cocaine	Higher rate of stillbirth, premature birth, low birth weight, learning difficulties; likely addiction to the drug.
Nutrition – folic acid	Good for the development of the brain and spinal cord; supplements should be taken as soon as pregnancy is recognised.

10.7

Pregnancy and baby

Getting pregnant | Pregnancy | Labour and birth | Your newborn | Babies and toddlers

Advice
1. Eat healthily. Eat more calories but get these from nutritious foods such as fruit and vegetables.
2. Limit alcohol intake and consider avoiding alcohol completely.
3. Do not smoke and try to avoid smoky places.
4. Take folic acid supplements.

FIGURE 1.10.7b: Some advice to pregnant women.

3. What are the common factors that badly affect the development of a foetus?
4. Look at Figure 1.10.7b. Explain why this advice is given to pregnant women. What advice can you give to pregnant mothers to help them have a healthy baby?

Validity and reliability in research

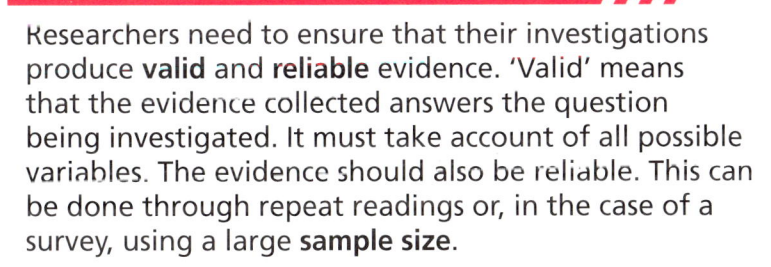

Researchers need to ensure that their investigations produce **valid** and **reliable** evidence. 'Valid' means that the evidence collected answers the question being investigated. It must take account of all possible variables. The evidence should also be reliable. This can be done through repeat readings or, in the case of a survey, using a large **sample size**.

5. Comment on the validity and reliability of the following studies:
   a) The first research on the effects of alcohol was conducted on 127 babies born to alcoholic mothers in France in 1968. The babies were found to have lower birth weights and lower intelligence.
   b) In a study on the effect of smoking, the ultrasound scans of 65 mothers who smoked were compared with the scans of 36 mothers who were non-smokers.

Did you know…?
A woman may not realise she is pregnant until about 8 weeks after conception (when the sperm and egg meet). The embryo's brain starts to develop after just 2 to 3 weeks and is highly affected by chemicals coming through the placenta.

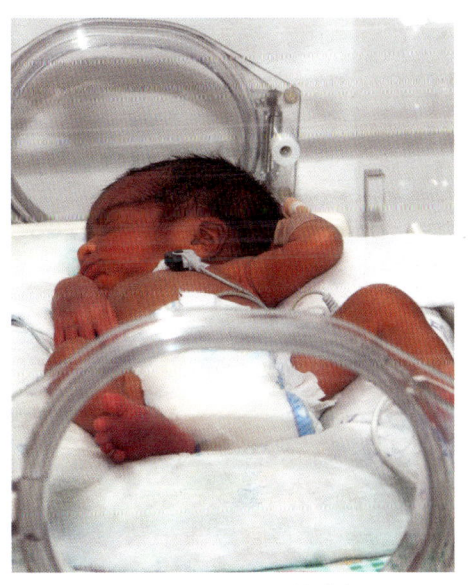

FIGURE 1.10.7c: A premature birth is a possible consequence of smoking during pregnancy.

Know this vocabulary
premature
valid
reliable
sample size

SEARCH: factors affecting a developing foetus 235

Genes

Communicating ideas about smoking in pregnancy

We are learning how to:
- Critique claims linked with the effects of smoking in pregnancy.
- Identify potential bias in sources of information.
- Give a reasoned opinion.

Nowadays, most people accept that smoking in pregnancy is harmful to the unborn baby. However, this is a fairly recent development. Why did it take so long for such an important message to be accepted?

Considering bias

Until the 1950s, cigarettes were not known to cause health issues. Even when evidence showed links with lung cancer of the smoker, it took much longer for the message about harm to an unborn baby to be accepted.

When evidence, or the conclusion from evidence, is swayed towards a certain outcome, we say it is biased. **Bias** may result from a mistake in an experimental procedure or can be caused on purpose, for example by someone who wants you to believe something.

1. Look at Figure 1.10.8b. Describe the message that is given to women by the advert.
2. Explain why there may be bias in the information in an advert.

FIGURE 1.10.8a: Some pregnant women continue to smoke, despite the compelling evidence that it is harmful to their unborn baby.

Critiquing a claim about smoking in pregnancy

A **claim** is a statement that says something is true. There must be facts or data that support a claim; these are the **evidence**. As more evidence was collected about the effects of smoking on the smoker, those around them and an unborn baby, more people's **opinions** started to change.

The more specific we can be in making a claim, the better. For example, 'smoking in pregnancy is harmful' is not as specific as 'smoking in pregnancy increases the risk of miscarriage'. The more specific the claim, the easier it is to demonstrate specific evidence to support the claim.

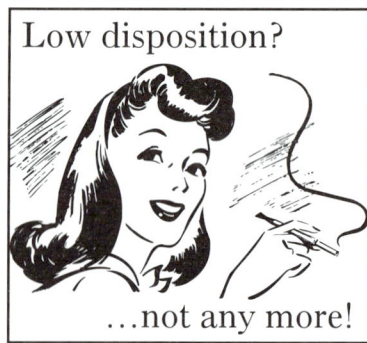

FIGURE 1.10.8b: Many 1950s adverts encouraged women to smoke and some even told them it was normal to feel nervous whilst pregnant and a cigarette could calm their nerves.

TABLE 1.10.8: Would this evidence convince you that smoking in pregnancy is harmful?

Claim	Evidence
Smoking in pregnancy increases the risk of premature birth.	Large-scale studies have shown that babies of smokers are approximately 15–200 g lighter at birth.
Smoking in pregnancy causes increased risk of the baby having cleft lip and palate.	Studies of 2000 babies showed that babies of mothers who smoked during pregnancy were 1.6 times more likely to be born with cleft lip and palate.
Smoking in pregnancy increases the risk of asthma in the baby.	Studies of 700 children showed that children of mothers who smoked more than 10 cigarettes per day in pregnancy were 2.5 times more likely to suffer from asthma.

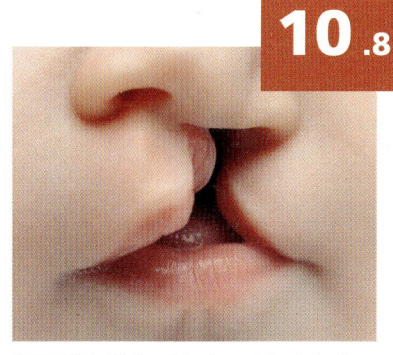

FIGURE 1.10.8c: A baby with cleft lip and palate.

3. Write a specific claim stating that there is a link between smoking in pregnancy and babies being born smaller than babies of non-smokers.

4. Suggest what evidence you would want to see to back up the claim in question 3.

Justify an opinion

We must be able to **justify** any opinion by explaining the ideas we have developed from the evidence. This is called **reasoning**. For example, we could have the opinion that the UK school day should be longer. We could use evidence such as the length of school day in various countries and a comparison of how their students do in tests. We could justify our opinion by explaining that the evidence shows that in China, for example, they have a longer school day and outperform the UK in some tests. Good justification would choose one or two other pieces of evidence to support the opinion. Finally, we should acknowledge other opinions and be prepared to defend our own opinion if someone disagrees with it.

It has been suggested that smoking in pregnancy should be banned. But some people feel that pregnant women should have free choice about whether they smoke or not.

5. In as much detail as possible:
   a) give your opinion on whether pregnant women should be banned from smoking;
   b) present the evidence that supports your opinion;
   c) explain the reasoning for your opinion;
   d) identify the opposite opinion to your own, and suggest how you would defend your opinion if someone disagreed with you.

Did you know…?

In 2015, a law came into force banning smoking in vehicles carrying anyone under the age of 18. Studies have shown that smoke can stay in the air for up to 2.5 hours after a cigarette is put out. This exposure to second-hand smoke is called passive smoking.

- State your opinion clearly
- List the facts, data and ideas that support your opinion
- Identify the most important piece of evidence and one or two other pieces of supporting evidence
- Explain how each piece of evidence supports your opinion
- Explain how you could defend your opinion to someone with a different viewpoint

FIGURE 1.10.8d: The stages in justifying an opinion.

Know this vocabulary

bias   opinion
claim   justify
evidence   reasoning

SEARCH: evidence about smoking in pregnancy

Genes

Checking your progress

To make good progress in understanding science you need to focus on these ideas and skills.

- ☐ Identify some features of different organisms of the same species.
- ☐ Explain the difference between continuous and discontinuous variation.
- ☐ Use data to explain whether variation is continuous or discontinuous and to investigate correlations between varying features.

- ☐ Identify examples of variation caused by inheritance and of variation caused by the environment in which the organism lives.
- ☐ Explain how a mix of genes from our parents means that siblings are different.
- ☐ Discuss the relationship between inherited features and the environment and describe how many features are caused by a combination, with examples.

- ☐ Recognise that variation within a species can help that species to survive.
- ☐ Use examples to describe how variation within a species can be an advantage if the environment changes.
- ☐ Make predictions about changes within a species to changes to external conditions.

- ☐ Name the main parts of the male human reproductive system.
- ☐ Describe the structures and functions of the main parts of the male human reproductive system; describe how fertility problems may arise.
- ☐ Explain how the male reproductive structures are designed for fertilisation; describe methods to combat infertility.

10.9

- Name the main parts of the female human reproductive system.
- Describe the structures and functions of the main parts of the female human reproductive system; describe how fertility problems may arise.
- Explain how the female reproductive structures are designed for fertilisation; describe methods to combat infertility.

- Recall the stages in development as a change from a single fertilised egg to an embryo and foetus.
- Compare the growth of the foetus at different stages. Describe the role of the mother in protecting the developing foetus.
- Describe the functions of different supporting structures of the mother.

- Identify substances passed on from a mother that will either help or harm her developing foetus.
- Describe how substances pass to and from a developing foetus and describe the effects of different factors on a developing foetus.
- Apply knowledge of effects of substances on advice given to pregnant women, considering validity of evidence.

- Identify bias in a claim and link it to claims about smoking in pregnancy.
- Explain what it means to critique a claim, and give examples of evidence to support a claim about the effects of smoking in pregnancy.
- Justify an opinion about smoking in pregnancy using evidence to support the opinion and to defend against an alternative opinion.

Genes

Questions

KNOW. Questions 1–6

See how well you have understood the ideas in the chapter.

1. An example of continuous variation is: [1]
   a) height;   b) tongue rolling;   c) blood type;   d) attachment of ear lobes.
2. State the two causes of variation. [2]
3. Explain why variation within a species is important. [4]
4. Which of the following will not pass from a mother to her developing foetus across the placenta? [1]
   a) Carbon dioxide   b) Carbon monoxide   c) Alcohol   d) Glucose
5. Male sperm cells are made in the: [1]
   a) penis;   b) testicles;   c) sperm duct;   d) urethra.
6. Outline what happens in the menstruation cycle. [4]

APPLY. Questions 7–12

See how well you can apply the ideas in this chapter to new situations.

7. Which of these features of butterflies is an example of discontinuous variation? [1]
   a) Area of wings   b) Length of body   c) Length of legs   d) Presence of spots
8. Suggest a variation of a feature that would be disadvantageous to a tiger. [1]
9. A rose seller plans to develop roses that grow in heavy, wet soil. Explain to the rose seller how this will affect variation in his roses and why it may not be a good idea. [2]
10. A woman is trying to get pregnant. Tell her the most likely time during her menstrual cycle to become pregnant. [1]
    a) Day 1   b) Day 5   c) Day 14   d) Day 21
11. A student midwife is explaining about the development of a foetus to some pregnant women. She has the diagram, right, and wants to show them where the baby grows. Which label should she choose? [1]

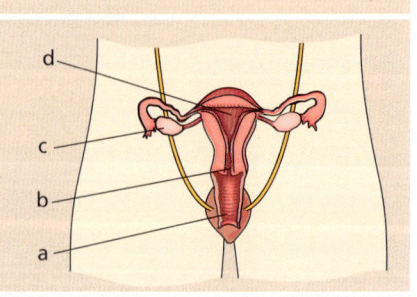

    FIGURE 1.10.10a

12. Two women are having treatment with drugs, one to increase fertility and the other to provide contraception. Explain how the drug each of them takes affects ovulation. [2]

10.10

EXTEND. Questions 13–15

See how well you can understand and explain new ideas and evidence.

13. Marvin is investigating shoe size among men. He measured the feet of 118 men. Figure 1.10.10b shows his results. [4]

    a) What is the sample size?
    b) What is the range?
    c) What type of variation is shown in the graph?
    d) What is the mode value of shoe size?

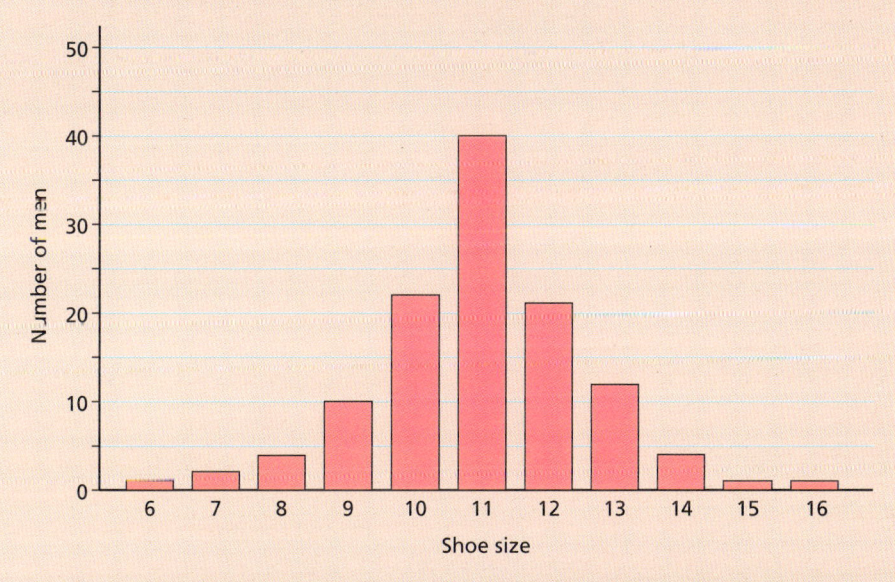

FIGURE 1.10.10b: Graph showing the results of an investigation into men's shoe size.

14. Scientists have carried out a survey to find out the effect of a diet pill on developing babies. Suggest one way that they can help to make evidence reliable. [1]

15. Sketch a graph to show how the weights of foetuses from smoking mothers compare to those from non smoking mothers over time. Give reasons for the differences. [4]

Glossary

absorption taking in, for example energy transferred by sound

accelerate speed up

acceleration how quickly speed increases or decreases

acid substance that has a pH lower than 7

air resistance frictional resistance when something moves through the air

alkali a soluble substance with a pH higher than 7

alloy mixture of two or more metals

ammeter used to measure the current flowing in a circuit

amniotic fluid liquid that surrounds and protects a foetus in the uterus

ampere unit of measurement of current, symbol A

amplitude maximum distance moved in a vibration, measured from the middle position

angle of incidence between the normal and incident ray

angle of reflection between the normal and reflected ray

antacid substance that neutralises stomach acid

antagonistic muscles pair two muscles that act in unison to create movement

anther pollen-producing part of the stamen of a flower

anticline an upfold in the Earth's crust

arthritis painful disease of the joints

attract pull towards; a magnet will attract any magnetic material that is close enough

auditory range the range of sound frequencies from the lowest to the highest that an animal or human can hear

average speed the overall distance travelled divided by the overall journey time

axis of rotation the centre line around which something rotates

bacteria (singular: bacterium) simple unicellular (single-celled) organisms, some of which can cause illness

balance when different elements of a system (physical, chemical, biological or ecological) are in equilibrium

base a substance that neutralises an acid; a base that dissolves in water is called an alkali

bias When an experimenter affects the outcome, or when a journalist favours a point of view

bicep upper muscle in the upper arm

bioaccumulation increase in the concentration of a chemical as it is passed from one organism to another up a food chain

boil when all of a liquid changes state to a gas, at the boiling point

boiling point the fixed temperature at which a pure liquid substance boils (or a gaseous substance condenses)

bone marrow tissue found inside some bones where new blood cells are made

brittle easily cracked or broken by hitting or bending

calcium hard mineral found in bone

carpel female part of a flower, made up of stigma, style and ovary

cartilage smooth tissue found at the end of bones, which reduces friction between them

cell 'building block' that all living things are made from

cell membrane layer around a cell that controls substances entering and leaving the cell

cell wall tough outer layer of plant cells, made of cellulose

Glossary

charge electrical energy that is positive or negative

charged up when some materials are rubbed together, electrons move from one surface to the other, leaving the surfaces electrically charged

chemical energy store energy store that is emptied during chemical reactions when energy is transferred to the surroundings

chemical reaction a change in which a new substance is formed

chloroplast structure in plant cells where light is absorbed so that photosynthesis can produce food

chromatogram pattern of results obtained in chromatography

chromatography process used to separate soluble substances

circulatory system the heart and blood vessels that transport essential substances around the body in blood

claim to present evidence and reasoning

combustion a reaction with oxygen in which energy is transferred to the surroundings as heat and light

competition struggle between different organisms for survival

component part of a complete unit (for example, in an electrical circuit)

compression Force squashing or pushing together

concave lens a lens that is thinner in the middle which spreads out light rays

concentration the number of particles in a certain volume of a substance

concentration gradient difference in concentration

conclusion opinion reached after studying all the evidence

condense change from a gas into a liquid when the temperature drops to the boiling point – as in water vapour condensing to liquid water

conduct transfer of heat or electrical charge by passing on energy to nearby particles

conductor (electrical) material that allows current to flow through it easily; it has a low resistance

consumer animal that eats other animals or plants

contact force a force that acts only if there is direct contact between objects

continuous variation where differences in characteristics (or other data) can have any numerical value, e.g. weight and height

control variable factor kept constant in an investigation

convex lens a lens that is thicker in the middle which bends light rays towards each other

core the centre of the Earth consists of a solid inner core and a liquid outer core

correlation how well sets of data are linked; high correlation shows that there is a strong link between two sets of data

corrosive reacts with materials and makes them dissolve

crust the rocky outer layer of the Earth

current rate of flow of electric charge, measured in amperes (A)

cytoplasm the jelly-like main component of cells, where most of the chemical processes occur

decibel (dB) unit of sound loudness

decomposer organism that breaks down dead plant or animal tissue

density mass of a material per unit volume

dependent variable variable that is measured in an investigation

deposition the geological process by which sediments, soil and rocks are added to a landform or land mass

diffusion the process in which particles in a liquid or gas move from an area of high concentration to an area of low concentration

Glossary

digestive system group of organs that together enable digestion of food, by breakdown and absorption of food molecules

discontinuous variation where differences in characteristics can only be grouped in discrete (separate) categories, e.g. eye colour, left- or right-handedness

dispersal distribution over an area (for example, of the seed away from the parent plant)

displacement reaction a chemical reaction in which a more reactive metal takes the place of a less reactive metal in a compound

dissipated when energy becomes spread out wastefully

dissolve when a solid mixes with a liquid so that it can no longer be seen

distance length of the path covered on a journey

distance–time graph a graph showing the relationship between distance and time

distillation process for separating liquids by evaporating then condensing the vapours

ductile able to be stretched out a lot

dull not good at reflecting light – opposite of shiny

echo reflection of a sound wave from a surface back to the listener

ecology study of the interactions between organisms and their environment

ecosystem the living things in a given area and their non-living environment

efficient a measure of how much of the energy transferred in a process achieved a desirable outcome. The greater the proportion of the output of a process that was useful, i.e. the smaller the proportion wasted, the more efficient it was

elastic energy store energy store filled when a material is stretched or compressed

electric current rate of flow of electric charge, measured in amperes (A)

electric field region around a charged object where another object feels an electrostatic force

electrical conductor material that allows current to flow through it easily; it has a low resistance

electrical insulator material that does not allow current to flow easily; has a high resistance

electron tiny negatively charged particle in an atom

electrostatic force non-contact force between two charged objects

embryo young foetus before its main organs are formed

energy something has energy if it has the ability to make something happen when the energy is transferred

energy resource a source of stored energy that can be released in a useful way

energy transfer diagram arrows that show how energy is transferred from one store to other stores

environment the surroundings, such as air, water, soil, climate, food sources, where an organism lives

equilibrium (in chemistry) a stable state, with no overall change

erosion the movement (transportation) of rock or soil by water, ice or wind

eukaryote one of a group of unicellular organisms that have a nucleus

evaporate change from a liquid to a gas at the surface of the liquid – such as when water evaporates to form water vapour

evidence information gathered in a scientific way, which supports or contradicts a conclusion

exoplanet planet that orbits a star outside our solar system

extinct when no more individuals of a species remain

extrusive the way that igneous rocks are formed at the surface of the Earth when hot magma is thrown from inside the Earth and then cools

fertilisation when the nucleus of a male sex cell fuses (joins with) that of a female sex cell

fertiliser chemical put on soil to increase soil fertility and allow better growth of crop plants

field (force field) an area where an object feels a force, e.g. area around a charged object where another object feels an electrostatic force

filament (of flower) 'stalk' of the stamen that supports the anther

filter material with microscopic holes used to remove insoluble solids from liquids

filtration separation of a solid from a liquid using a filter

fissure a fine, long crack in the Earth's surface

foetus developing baby during pregnancy

food chain part of a food web, starting with a producer, ending with a top predator

food security availability of food and the ability to obtain it

food web flow diagram showing how a number of living things in a habitat get their food

force a push, pull or turning effect, measured in newtons (N)

formula (in chemistry) chemical symbols and numbers that show which elements, and how many atoms of each, a compound is made up of

formula (in physics) equations that show the relationships between different properties

fossil the preserved remains of a living thing that lived millions of years ago

fossil fuel coal, natural gas and crude oil that were formed from the compressed remains of ancient plants and other organisms

fracture broken bone

free electron negatively charged particle that moves freely within a metal

freeze change from a liquid to a solid when the temperature drops to the melting point

freeze–thaw a process that weathers rocks – water seeps into tiny cracks in rock, the water then freezes and expands as it forms ice, which causes the rock to crack further and pieces may break off

frequency number of events in unit time, for example, number of waves produced by a source in one second; unit hertz (Hz)

fruit ovary of a plant after fertilisation; it contains seeds

fuel material that is burned to release its energy

galaxy a group of billions of stars held together by gravity

gamete the male gamete (sex cell) in animals is a sperm, the female gamete is an egg

gas pressure caused by collisions of particles with the walls of a container

genetic caused by genes, inherited

germination when a seed begins to grow into a plant

gestation process where the baby develops during pregnancy

gravitational field region in which objects exert an attractive force on each other (dependent on their mass)

gravitational field strength, g the gravitational force on unit mass (1 kg) at a particular point; measured in newtons per kilogram, N/kg

gravitational potential energy store energy store that is filled when an object is raised

gravity force that pulls masses towards one another, e.g. an object towards the Earth

hardness a measure of how easy it is to scratch a solid

Glossary

Glossary

hearing range range of sound frequencies that an animal can hear

hertz (Hz) unit of frequency, equal to one per second

hydrogen type of atom present in all acids; hydrogen gas is formed when a metal reacts with an acid

hydroxide particle present in all alkalis

hypothesis an explanation you can test which includes a reason and a 'science idea'

igneous rocks rocks formed from the solidification of cooled magma, with minerals arranged in crystals

image picture of an object that we see in a mirror or through a lens or system of lenses

immiscible liquids that do not mix, but form separate layers

immune system the parts of the body that protect it against infections by pathogens such as viruses and bacteria; includes the white blood cells

incident ray the incoming ray

independent variable a variable in an experiment that affects the outcome

indicator chemical that is a different colour in an alkali and in an acid, used to identify whether an unknown solution is acidic or alkaline

infertility inability to reproduce by natural methods

inherited a feature or characteristic that has been passed on from parent to offspring, genetically

insecticide chemical applied to crops to destroy insects that damage crops

insoluble unable to dissolve

insulator (electrical) material that does not allow current to flow easily; has a high resistance

interdependence mutual reliance between two or more groups. In an ecosystem populations of different organisms may affect each other

intermolecular forces forces between molecules

intrusive igneous rock formed from magma forced into older rocks within the Earth's crust

irritant something that irritates and reddens the skin

joint where two bones meet; allows movement

joule unit of energy, symbol J

justify the process of proving that an idea is correct or incorrect

kilojoule unit of energy equal to 1000 J, symbol kJ

kilopascal standard unit of pressure

kilowatt unit of power equal to 1000 watts or joules per second

kilowatt-hour (kWh) the energy transferred in 1 hour by an electrical appliance with a power rating of 1 kW

kinetic energy store energy store filled when a moving object speeds up

lava molten rock (magma) from beneath the Earth's surface that has erupted from a volcano onto the Earth's surface

lens a specially shaped piece of transparent material that refracts light passing through it to form an image

ligament connects bone to bone; made of stretchy fibres called collagen

light year (ly) the distance travelled by light in one year

lithosphere the rocky outer section of the Earth, consisting of the crust and upper part of the mantle

litmus indicator solution

longitudinal wave wave in which the vibrations are parallel to the direction in which energy is transferred

magma molten rock inside the Earth

magnetic material that is attracted by a magnet

magnification a measure of how many times bigger an image is than the object

Glossary

mains supply household alternating current electric power supply

malleable able to be bent without breaking

mantle the semi-liquid layer of the Earth beneath the crust

mass the amount of matter (stuff) in an object, measured in kilograms (kg)

mean average

melt change from solid to liquid when the temperature rises to the melting point

melting point the fixed temperature at which a pure solid substance melts (or a liquid substance freezes)

menstruation cyclical breakdown of the uterus lining, leading to bleeding from the vagina (a period)

metal shiny, good conductor of electricity and heat, malleable and ductile, and usually solid at room temperature

metamorphic rocks rocks formed from existing rocks exposed to heat and pressure over a long time

metamorphose change completely

microscope optical device used to see magnified images of tiny objects and structures

minerals chemicals from which rocks are made

mitochondria structures in a cell that produce energy

mixture two or more elements or compounds mixed together, but not chemically joined

model something which explains an aspect or part of the physical world

molecule two to thousands of atoms joined together. Most non-metals exist either as small or giant molecules

monoculture growing of a single crop in an area

multicellular living thing made up of many types of cell

muscle soft tissue with filaments that slide past each other, contracting to produce force, e.g. motion, heartbeat or peristalsis to move partly digested food

muscular skeletal system the bones of the skeleton, the ligaments, the skeletal muscles and the tendons all working together

negatively charged an object that has gained electrons as a result of a charging process

nervous system network of nerve cells and fibres in the body that carry electrical impulses around the body

neutral (in chemistry) when a substance is neither acidic nor alkaline

neutralisation a chemical reaction in which an acid and a base react with each other resulting in a neutral solution, i.e. one that is neither acid or alkaline

neutron star the product of the explosive transformation of a massive star

newtons unit for measuring force

non-contact force a force that acts without direct contact between objects

non-metal dull, poor conductor of electricity and heat, brittle and usually solid or gaseous at room temperature

non-renewable (resource) an energy source that will be used up, e.g. fossil fuel

normal line from which angles are measured, at right angles to the surface

nuclear fusion a reaction in which two or more atomic nuclei come close enough to form one (or more) different nuclei and other atomic particles, releasing large amounts of energy

nucleus (of cell) part of a cell that contains the genetic material (DNA) which controls the cell's activities

Glossary

ohm unit of measurement of electrical resistance, symbol Ω

opaque material that allows no light to pass through

opinion a view formed about something, not necessarily based on fact or supported by evidence

orbit the path taken by a satellite, planet or star moving around a larger body. Earth completes one orbit of the Sun every year

organ collection of tissues that work together to perform a function

organ system organs that coordinate with one another in body processes

organism a living thing

oscilloscope device that allows sound waves, which have been turned into electrical signals, to be viewed as waveforms

osteoporosis disease in which the density of bones drops below a healthy level and bones become fragile, making them prone to fractures

ovary organ in female animals that makes egg cells; and in plants that contains ovules

oviduct tube in a female animal that carries the egg cell from the ovary to the uterus, and where fertilisation occurs

ovulation release of an egg cell from the ovary during the menstrual cycle

ovule female sex cell of a plant

oxidation chemical reaction in which a substance combines with oxygen

parallel circuit electric circuit in which each component is connected separately in its own loop

particle very small part of a material, such as an atom or a molecule

particle model a way of explaining the behaviour of solids, liquids and gases, in terms of the small moving particles in the substance

penis sex organ of a male animal, which carries sperm out of the body

pesticide chemical applied to crops to destroy pests

pH measure of acidity/alkalinity, on a scale from 0 to 14

pitch how high or low the frequency of a sound is

placenta organ that provides the foetus with oxygen and nutrients and removes waste substances

pollen grain that contains the male sex cell of a flower

pollen tube structure that grows from a pollen grain in order to fertilise an ovule

pollination process of transferring pollen from the anther of a flower to the stigma of a flower

population the number of a type of organism living in a particular area

positively charged an object that has lost electrons as a result of a charging process

potential difference amount of energy shifted in an electric circuit (from battery to charge or from charge to component) per unit charge; also called voltage, unit volt (V)

power amount of energy that something transfers each second, measured in watts (W)

predator animal that preys on (and eats) another animal

premature when a baby is born before it has fully developed

prey an animal that is hunted and killed by other animals

producer component of a food web or chain that produces its own food (typically a green plant)

prokaryote one of a group of unicellular organisms that have no nucleus

proton positively charged particle in the nucleus of an atom

pure (substance) containing only one type of element or compound

purify make pure, or nearer to pure

Glossary

ray model a simple diagrammatic model used to show how light behaves as it reflects off a mirror or passes through transparent materials

reactivity the tendency of a substance to undergo a chemical reaction

reactivity series the ordering of metals by how vigorously they react, with the most reactive first

reasoning the act of thinking about something in a logical way; the steps can be used to justify the conclusion reached

recreational drugs use of drugs by a person to alter their emotions, perceptions or feelings for recreational purposes

red giant an old star that has expanded greatly and appears red

reflected ray the outgoing ray after reflection from a surface

reflection when a wave, such as sound or light, bounces off a surface

refraction a change in the direction of a wave such as light, caused when it enters a material of a different density

relative motion the motion of an object as seen by an observer in motion

relative speed the speed of an object as calculated by an observer in motion; it depends on the observer's speed

reliable results from an experiment which display overall consistency; it produces similar results under consistent conditions

renewable (resource) an energy resource that will not run out, e.g. solar energy and wind energy

repeatable when repeat readings are close together

repel push away; two similarly charged objects will repel one another

reproductive system organs in a male or female organism involved in producing offspring; in humans it is where sperm or eggs are produced

resistance (in electricity) property of an electrical component, making it difficult for charge to pass through; unit of measurement is the ohm (Ω)

respiratory system organ system in living things in which oxygen is taken in and carbon dioxide is removed

retina back layer of the eye, with light-detecting cells where an image is formed

ring main how the electricity supply in a house is connected

rock cycle the relationships between different types of rock and the processes that occur to change these over long periods of time

salt type of chemical compound – our table salt is sodium chloride

sample size the number of observations to include in a sample as part of an investigation

Sankey diagram energy transfer diagram that shows the proportion of energy transferred in different ways

scattering when light bounces off an object in all directions

season a division of the year, marked by changes in weather, ecology and hours of daylight; summer, autumn, winter and spring

sedimentary rocks rocks formed from layers of sediment; can contain fossils

seed structure that contains the embryo of a new plant

semen fluid in which sperm are carried

series circuit electric circuit in which all components are connected one after the other in the same loop

skeletal system *see muscular skeletal system*

skeleton all the bones in the body

solubility the mass of solute that dissolves in a solvent at a particular temperature

soluble solid that can dissolve (usually in water)

solute solid that has been dissolved

Glossary

solution mixture formed when a solid dissolves in a liquid

solvent liquid in which something dissolves

sonorous make a sound like a bell when hit

soundproofing using materials that absorb sound

specialised cell cells that have adapted to fulfil a specific function

species group of organisms that have more in common with each other than with other groups; they can interbreed and produce fertile offspring

spectrum a wide range of values, for example of frequencies or wavelengths in the visible spectrum of light

speed how fast something travels – how much distance is covered in how much time

sperm duct tube through which sperm travel from the testes

stamen male part of a flower, made up of filament and anther

star massive, luminous sphere of plasma held together by its own gravity

static electricity an imbalance of electric charges on the surface of a material

stationary still, not moving

stigma pollen-receiving part of a flower

strata layers of sedimentary rock

strength ability of a solid to withstand a force

structural adaptations special features, for example of a cell, to enable specific functions to be carried out

style (of flower) female part of a flower through which pollen travels to fertilise an ovule

sublime process where a solid changes directly into a gas; there is no liquid state

survival advantage variations in individuals that give the species more chances to survive

syncline bowl-shaped layer of rock, where the rock has been forced downwards

tectonic plate a section of the Earth's crust that slowly moves relative to other plates

tendon connects muscle to bone; made of stretchy fibres called collagen

testicle organ of a male animal where sperm are made

thermal energy store energy store filled when an object is warmed up

time-lapse sequence a series of images are recorded and then played back at a higher frequency

tissue collection of body cells of one type that work together to carry out a task

titration process to find out how much of a chemical there is in a solution by addition of another (liquid) chemical of known strength until an endpoint is reached

toxin substance that damages a living organism

translucent a material that lets some but not all light pass through

transparent a material that allows light to pass through

tricep lower muscle in upper arm

trophic level position of an organism in a food chain

umbilical cord tissue that connects a foetus to its mother's placenta

unicellular living thing made up of just one cell

unit standard amount used to measure a physical quantity – for example metre, kilogram and second

universal indicator turns a range of different colours depending on how strongly a solution is alkaline or acidic

upfold a dome-shaped layer of rock that has been pushed upwards

uplift upwards movement of the Earth's surface in response to natural processes such as earthquakes

Glossary

urethra in a male, a tube in the penis through which sperm travel in semen

uterus part of a woman's body where a foetus develops – also called the womb

vacuole part of a cell that contains liquid, and can be used by plants to keep the cell rigid and store substances

vacuum a space where there are no particles of matter

vagina part of a female body where the penis enters and sperm is received

valid the suitability of a procedure to answer a particular question

variable factor that may affect the outcome of an experiment

vapour liquid that has evaporated

variation differences in characteristics between individuals of the same species and between species

vibration to-and-fro movement that repeats

viscosity resistance to flow of a liquid; a viscous liquid is slow-flowing

volt unit of measurement of voltage (potential difference)

voltage measure of the size of 'push' that causes a current to flow around a circuit; it is the amount of energy shifted (from battery to charge, or from charge to component) per unit charge

voltmeter device used to measure the voltage across a component in an electric circuit

volume measurement of amount of space a material takes up, unit cm^3 or m^3; also a measurement of how loud a sound is, unit decibel (dB)

watt unit of power, or rate of transfer of energy, equal to a joule per second; symbol W

waveform graph of the displacement of a wave motion, at different distances along the wave (or at different times)

wavelength distance along a wave from one point to the next corresponding point where the wave motion begins to repeat itself – for example crest to crest

weathering the breaking down of rocks, soil and minerals, by physical, chemical or biological processes

weight force of gravity acting on an object, measured in newtons (N)

Index

absorption 86, 87, 97
acceleration 10, 12-3, 14–15
    distance-time graphs 10–11, 12–13, 16
    in gravitational fields 21
acids 136–7, 141
    reactions with alkalis 142–3
    reactions with metals 130–31
adaptation 188, 220
air resistance 18, 21
alkalis 138–9, 141
    reactions with acids 142–3
alloys 107, 128–9
ammeters 32–3
amniotic fluid 233
amperes (A) 33
amplitude 79, 80–1
angle of incidence 91
angle of reflection 91
animal cells 186, 187, 188
antacids 144
antagonistic muscle pairs 180
anthers 208
anticlines 161
arthritis 183
atoms 44
attraction 42, 46
auditory range 82
average speed 9, 11
axis of rotation 164, 165

bacteria 187, 192–3
    antibiotic resistance of 227
balanced forces 19
bases 135, 144
batteries see cells (electrical)
bias 236
bicep 179, 180
bioaccumulation 203
blood 115, 177, 184–5
boiling 112, 113, 118, 119
boiling point 113
bone marrow 177
bones 176–7, 178–9, 182–3
brittle materials 106
burning 55, 60, 66

calcium 177
carpels 208
cartilage 179
cell division 232

cell membrane 186–187
cell wall 187
cells (electrical) 32, 34
cells (in organisms) 186–91, 232
changes of state 112–13, 119
charge 42–7
chemical energy stores 64
chemical reactions 130–35
chloroplasts 187, 189, 192–3
chromatograms 120–1
chromatography 120–21
circulatory system 184–5
claims 236–7
classification 220
combustion 135
    see also burning
competition 206–7
components in circuits 32, 34–5
compression 108–9
concave lenses 93, 95
concentration 103, 110–111, 137
concentration gradient 110, 111
condensation 112, 113, 118, 119
conduction 36, 106, 128
    see also electrical conductors
consumers 201, 203
contact forces 18, 43
continuous data 15
continuous variation 222–3
control variables 15, 144
convex lenses 93, 95
correlation 15, 144, 223
corrosive substances 136–7, 138
crust 152, 153
crystals 116–7, 131, 150, 154-5, 158
current 32, 33–8, 42
    in parallel circuits 38–9, 40
    in series circuits 38–9, 40
cytoplasm 186–7

data, presentation 15
day 164, 168
decibels (dB) 79
decomposers 200–1
density 105
dependent variables 15, 144
deposition 156, 160
diffuse reflection 90–91
diffusion 110–111, 193
digestive system 185

Index

discontinuous variation 222–3
discrete data 15
displacement reactions 132–3
dissipation of energy 62, 64
dissolving 116
distance 8–9, 166–7
distance-time graphs 10–13
distillation 118–19
DNA fingerprinting 121
drugs, effect on foetus 234
ductile materials 106, 128

ears 78–9, 82–3
Earth
    gravitational field 6, 7, 20–21, 23, 69
    movement of 164–5, 168
    structure of 152–3
echoes 86
eclipses 89
ecology 206
ecosystems 206
elastic energy stores 65, 70–71
electric circuits 32–3
    energy in 34–5
    parallel 38–41
    series 38–41
electric fields 42–3, 46–7
electrical conductors 32, 36–7, 42, 106, 107, 128
electrical insulators 32, 36–7
electricity
    generation 60–61
    in the home 41
    responsible use 62–3
electrons 44
    charge and 44–7
    current and 32, 33, 36–7
electrostatic force 42–3
embryos 232–3
energy
    in circuits 34–5
    conservation of 69
    cost 59, 62–3
    dissipation 62, 64
    in food 54–5
    from fuels 54
    in the home 54, 58–9
    in particles 104
    in sound waves 78–9, 80
    sources 60–61, 66
    storage 55, 64–5, 70
    transfer 55–7, 64–9
energy transfer diagrams 66–7
environment
    electricity generation and 61, 63
    organisms' effect on 206–7
    toxins in 202–3
    variation and 224
equilibrium 110
erosion 156, 160
eukaryotes 192, 193
evaporation 113, 119
evidence 235, 236–7
exoplanets 163
extinction 226
extrusive rock 154–5, 160
eyes 94–5, 180

faulting 161
female reproductive system 228–9
fertilisation
    in humans 231
    in plants 210–211
fertilisers 202
fields 20
    electric 42–3, 46–7
    gravitational 20–21, 23, 69
filaments 208
filters 114–5
filtration 115
fissures 155
flowering plants 208–211
foetus 228, 231–5
folding 161
food 54–5
food chains 200–1
food security 204–5
food webs 200–201
forces 18, 181
    in balance 19
    electrostatic 31, 42–3
    intermolecular 105, 107, 108
    resultant 19
formulae 9
fossil fuels 60
fossils 156–7, 159
fractures 182, 183
free electrons 36–7
freeze-thaw weathering 157
freezing 112, 113
frequency
    light waves 96–7
    sound waves 80–81, 82
fruits 211, 214
fuel bills 58
fuels 54, 55, 60–61

galaxies 163
gametes 231
gas pressure 109
gases 104, 108–9, 110

Index

particle model 104, 105, 109
    sound in 84–5
genetic features 224–5
germination 212–13
gestation 232–3
gravitational field strength 20–21, 23, 69
gravitational fields 20–21, 23, 69
gravitational potential energy 64–5, 68–9
gravity
    as a force 18, 20, 22–5, 68–9
    in space 19, 23, 24–5
guard cells 189

hardness 106
hearing 82–3
heart 178, 184–5
hertz (Hz) 80
hormones 228–9, 230
hydrogen 130–1, 137
hydroxide particles 138–9

igneous rocks 154–5, 160, 169
images 90
immiscible liquids 115
immune system 185
incident rays 90–1
independent variables 15, 144
indicators 140–41, 142
indigestion remedies 138, 144–5
infertility 229
inherited features 224–5
insecticides 202–3
insects 204–5, 208–9
insoluble substances 115
interdependence 206–7
intermolecular forces 105, 107, 108, 113
intrusive rock 154–5
irritants 136–8

joints 177, 178, 179, 180, 183
joules (J) 54, 57
justification 237

kilojoules (kJ) 54
kilopascals (kPa) 109
kilowatt-hours (kWh) 58
kilowatts (kW) 56
kinetic energy stores 65, 68

lava 153, 154–5, 161
lenses 92–3, 94–5
ligaments 178
light
    coloured 96–7
    properties 88–9

reflection 90–91
refraction 90-1, 92–3, 96
speed of 166
as waves 90, 92, 96–7
light bulbs 32, 62–3, 66–7
light years 166
liquids 104, 108–111
    particle model 104, 105, 109
    sound in 84–5
lithosphere 152, 153
litmus 140, 141
longitudinal waves 78–9

magma 153, 154–5, 160
magnetic materials 114, 128
magnification 191
mains supply 41
male reproductive system 230–31
malleable materials 106, 128
mantle 152–3
mass 20–1, 22–3
mean values 145
melting 112, 113
melting point 106, 113, 128
menstruation 229, 231
metals
    as conductors 36–7, 42, 128
    properties 105, 106–7, 128
    reactions with acids 130–31
metamorphic rocks 158–9, 160, 169
metamorphosing 158
microphones 81
microscopes 93, 190–91
minerals 157, 158–9
mitochondria 186, 187
mixtures 114–15, 118–19, 120-1
models 168–9
    day length 168
    electric circuits 33, 35, 36
    energy storage and transfer 64–5
    light 90, 92
    particle 84–5, 104–7, 109, 110
    rocks 169
    seed dispersal 213
monoculture 205
Moon
    eclipses and 89
    gravity and 20, 24–5
    movement of 167
multicellular organisms 184–5
muscle cells 184, 188
muscles 178–81
muscular skeletal system 178–83, 185

negatively charged 44–5

Index

nerve cells 188
nervous system 185
neutral solutions 140
neutralisation reactions 142–5
neutron stars 162
Newtons (N) 20, 22, 181
night 164
non-contact forces 20, 43
non-metals 106, 129
non-renewable energy sources 60–61
nuclear fusion 162
nucleus 44, 186–7

ohms (Ω) 37
opaque materials 88, 97
opinions 236–7
orbit 19, 23, 24
organ systems 184–5
organisms 184–5
   effect on environment 206–7
organs 175, 184
oscilloscopes 81
osteoporosis 183
ovaries
   human 228
   plant 208, 210, 211
oviduct 228–9
ovulation 229, 231
ovule 208, 210, 211
oxidation 134–5

parallel circuits 38–41
particle model 84–5, 104–7, 109, 110–1
particles 104
penis 230–1
pesticides 202–4
pH scale 137, 139, 140–41
pitch 80
placenta 233, 234
plants
   plant cells 186, 187, 189
   reproduction in 208–211
plasma 105
pollen 208–11
pollen tube 210–1
pollination 204–5, 208–210
population 200, 206
positively charged 44–5
potential difference 35, 37, 39
   see also voltage
power 53, 56–7
predators 203, 206, 207
premature birth 234–5
pressure 109
prey 201, 207

producers 201
prokaryotes 192–3
protons 44
puberty 221
pure substances 113–4
purification 119

ramps, motion on 14–15
ray models 90, 92
reactivity 130–1
reactivity series 132–3
reasoning 237
red giant 162–163
reflected rays 90–1
reflection
   light 90–91
   sound 86–7
refraction 90, 92–3, 96
relative motion 16–17
relative speed 16–17
reliable evidence 235
renewable energy sources 60–61
repeatable experiments 145
reproductive systems
   human 185, 228–31
   plants 208–211
repulsion 42–3, 46
resistance 36–9
respiratory system 185
resultant force 19
retina 94, 95
reversible changes 112
ring mains 40, 41
rock cycle 160–61
rocks 154–61, 169
root hair cells 189
rusting 134

salts 130–33, 143
sample size 223, 235
sampling 214–15
Sankey diagrams 67
scattering 90
seasons 164–5, 168–9
sedimentary rocks 156–7, 160–1, 169
seeds
   development 211
   dispersal 212–15
semen 230
separation of mixtures 114–15, 118–19, 120-121
series circuits 38–39, 40–41
shadows 88, 89
skeleton 176–7
smoking 229, 234, 236–7
soils 150, 202

255

Index

solids 106–7
    particles in 104, 105, 106–7
    sound in 84–5
solubility 106, 107, 116–17
soluble substances 106, 115–17
solutes 116, 117
solutions 116–17
solvents 116, 117
sonorous materials 128
sound 78–81
    absorption 86–7
    in materials 84–5
    properties 80–81
    reflection 86–7
    speed of 17, 84
    in a vacuum 84
soundproofing 87
specialised cells 188–9, 232
species 222, 225, 226
spectrum 96–7
specular reflection 88, 90–91
speed
    average 9, 11
    distance and 8–9
    relative 16–17
speed cameras 8, 9, 12
sperm cells 188, 224, 230–31
sperm duct 230
stamens 208, 211
stars 23, 24, 150, 162–3, 166–7
states of matter 102, 104, 110
static electricity 42–7
stationary objects 11
stem cells 188–9, 232
stigma 208–211
strata 156–7, 161
strength
    materials 106
    muscles 181
structural adaptations 188
style 208, 210
sublimation 112
Sun 105, 150, 162
    Earth's rotation and 150, 151, 164–5, 168
    eclipses and 89
    energy from 66, 67, 187
survival advantage 226
synclines 161

tectonic plates 152, 153, 160
temperature
    particle model and 103, 104, 109, 110
    solubility and 117
tendons 178

testicles 230
thermal energy stores 64–5
time-lapse sequences 12–13
tissues 175, 184
titration 142–3
toxins 202–3
translucent materials 88
transparent materials 88, 92, 97
tricep 180
trophic levels 200–1

umbilical cord 233
unicellular organisms 186, 192–3
units 8, 56, 58, 109, 166
universal indicator 140, 141, 142
upfolds 161
uplift 160, 161
urethra 230
uterus 221, 228

vacuole 187
vacuum, waves in 77, 84, 96
vagina 228
valid evidence 235
van de Graaff generators 45, 47
vapour 103, 118
variables 15, 144
variation 220–1, 222–7
vertebrates 174, 176
vibrations 52, 76, 78
viscosity 108, 109
volcanoes 151, 153, 154–5
voltage 30, 31, 34–8
    in parallel circuits 31, 40
    in series circuits 31, 40
voltmeter 34, 35
volts (V) 34
volume 79

watts 56, 59, 72
waveforms 77, 81
wavelength 77, 80, 81, 96–7
waves
    energy in 76–7, 80
    light 52, 77, 90, 92, 96–7
    sound 52, 77, 78–81
    in a vacuum 77, 84
weathering 151, 156, 157–8, 160
weight 7, 20–4

Notes

Notes

Notes

Acknowledgements

The publishers wish to thank the following for permission to reproduce photographs. Every effort has been made to trace copyright holders and to obtain their permission for the use of copyright materials. The publishers will gladly receive any information enabling them to rectify any error or omission at the first opportunity.

(t = top, c = centre, b = bottom, r = right, l = left)

Cover and title page images: arigato/Shutterstock, (tl) Artem Kovalenco/Shutterstock, (tr) fuyu liu/Shutterstock, (c) NikoNomad/Shutterstock, (bl) Pavel Vakhrushev/Shutterstock, (bc) robert_s/Shutterstock, (br) Sailorr/Shutterstock

p6–7 Omikron/Science Photo Library, p6 (t) Aflo Co. Ltd./Alamy Stock Photo, p6 (c) Vitalii Nesterchuk/Shutterstock, p6 (b) Keith Publicover/Shutterstock, p7 (t) Christian Mueller/Shutterstock, p7 (b) NASA, p7 (c) Markus Mainka/Shutterstock, p8 (t) Tom Gowanlock/Shutterstock, p8 (b) Bryon Palmer/Shutterstock, p9 Aflo Co. Ltd./Alamy Stock Photo, p10 Spotmatik Ltd/Shutterstock, p12 Brian A Jackson/Shutterstock, p13 Juan J. Jimenez/Shutterstock, p16 Christian Mueller/Shutterstock, p17 (t) Dmitrydesign/Shutterstock, p17 (b) marigo20/Shutterstock, p18 (t) Vitalii Nesterchuk/Shutterstock, p18 (b) Markus Mainka/Shutterstock, p19 Pressmaster/Shutterstock, p20 (t) Adrian Hughes/Shutterstock, p20 (b) Gordon Bell/Shutterstock, p21 (t) Dorling Kindersley/Getty Images p21 (b) Erich Schrempp/Science Photo Library, p22 (t) Kletr/Shutterstock, p22 (b) Dimitar Sotirov/Shutterstock, p23 NASA, p24 Keith Publicover/Shutterstock, p30–31 Pi-Lens/Shutterstock, p30 (t) GIPhotoStock/Science Photo Library, p30 (b) Photo Melon/Shutterstock, p30 (c) GIPhotoStock/Science Photo Library, p31 (t) Trever Clifford Photography/Science Photo Library, p31 (ct) Science Photo Library/Science Photo Library, p31 (cb) Science Photo Library/Science Photo Library, p31 (b) GIPhotoStock/Science Photo Library, p33 (l) abutyrin/Shutterstock, p33 (r) Mostovyi Sergii Igorevich/Shutterstock, p34 sciencephotos/Alamy Stock Photo, p35 Lily81/Shutterstock, p39 Matthew Gough/Shutterstock, p41 Adrian Britton/Shutterstock, p42 (t) Ted Kinsman/Science Photo Library, p42 (b) GIPhotoStock/Science Photo Library, p44 Friedrich Saurer/Science Photo Library, p46 Martyn F. Chillmaid/Science Photo Library, p47 Adam Hart-Davis/Science Photo Library, p52–53 Luis Abrantes/Shutterstock, p52 (t) sgm/Shutterstock, p52 (c) Benjamin Marin Rubio/Shutterstock, p52 (b) Hurst Photo/Shutterstock, p53 (t) Santhosh Varghese/Shutterstock, p53 (b) Cordelia Molloy/Science Photo Library, p54 Jenoche/Shutterstock, p56 (t) Gaus Nataliya/Shutterstock, p56 (ct) Ami Parikh/Shutterstock, p56 (cbl) Jiang Hongyan/Shutterstock, p56 (bc) Hurst Photo/Shutterstock, p56 (cbr) ppart/Shutterstock, p56 (bl) You can more/Shutterstock, p56 (br) sutsaiy/Shutterstock, p57 Sheila Terry/Science Photo Library, p59 (t) Deyan Georgiev/Shutterstock, p59 (b) sgm/Shutterstock, p62 Awe Inspiring Images/Shutterstock, p63 Jezper/Shutterstock, p64 Claudia Paulussen/Shutterstock, p65 (t) Ulga/Shutterstock, p65 (c) Santhosh Varghese/Shutterstock, p65 (b) CroMary/Shutterstock, p66 (t) Cordelia Molloy/Shutterstock, p66 (bl) Neamov/Shutterstock, p66 (br) Neamov/Shutterstock, p68 (t) Kotsovolos Panagiotis/Shutterstock, p68 (b) Germanskydiver/Shutterstock, p70 (t) Ian Cumming/Science Photo Library, p70 (c) 2happy/Shutterstock, p70 (b) bunnyphoto/Shutterstock, p71 (t) Mauro Rodrigues/Shutterstock, p71 (b) Africa Studio/Shutterstock, p76–77 Andrew Lambert Photography/Science Photo Library, p76 (t) Ysbrand Cosijn/Shutterstock, p76 (ct) Vladimir Gjorgiev/Shutterstock, p76 (cb) Alex Hubenov/Shutterstock, p76 (b) rangizzz/Shutterstock, p77 (t) michaeljung/Shutterstock, p77 (b) Kukhmar/Shutterstock, p78 Ysbrand Cosijn/Shutterstock, p79 Juergen Freund/Alamy Stock Photo, p80 CRSHELARE/Shutterstock, p82 Tsekhmister/Shutterstock, p84 Ethan Daniels/Shutterstock, p86 (t) michaeljung/Shutterstock, p86 (b) BlueRingMedia/Shutterstock, p87 (l) BlueRingMedia/Shutterstock, p87 (r) Raphael Daniaud/Shutterstock, p88 (l) Alex Hubenov/Shutterstock, p88 (r) auremar/Shutterstock, p89 (l) Igor Kovalchuk/Shutterstock, p89 (b) Marius Meyer/Shutterstock, p90 (t) Kukhmar/Shutterstock, p90 (b) Samuel Borges Photography/Shutterstock, p91 (t) Sergey Krasnoshchokov/Shutterstock, p91 (b) BlueRingMedia/Shutterstock, p92 Peter Sobolev/Shutterstock, p96 GIPhotoStock X/Alamy Stock Photo, p102–103 Lawrence Lawry/Science Photo Library, p102 (t) Achim Baque/Shutterstock, p102 (c) Andrew Lambert Photography/Science Photo Library, p102 (b) Patricia Hofmeester/Shutterstock, p103 (t) Clive Freeman/Biosym Technologies/Science Photo Library, p103 (b) Charles D. Winters/Science Photo Library, p105 Achim Baque/Shutterstock, p106 Africa Studio/Shutterstock, p108 (t) Olga Miltsova/Shutterstock, p108 (b) Oliver Hoffmann/Shutterstock, p109 Jahthanyapat/Shutterstock, p110 Africa Studio/Shutterstock, p111 Andrew Lambert Photography/Science Photo Library, p112 phloen/Shutterstock, p113 (l) WhiteTag/Shutterstock, p113 (r) Steven Coling/Shutterstock, p114 Charles D. Winters/Science Photo Library, p116 (t) Dario Lo Presti/Shutterstock, p116 (b) Javier Trueba/MSF/Science Photo Library, p118 Fotocrisis/Shutterstock, p119 (tl) Tetiana Yurchenko/Shutterstock, p119 (tr) tanuha2001/Shutterstock, p119 (ctl) givaga/Shutterstock, p119 (ctr) M. Unal Ozmen/Shutterstock, p119 (cbl) Ilya Akinshin/Shutterstock, p119 (cbr) Vitaly Korovin/Shutterstock, p119 (bl) xpixel/Shutterstock, p119 (br) Fablok/Shutterstock, p120 Charles D. Winters/Science Photo Library, p121 (l) isak55/Shutterstock, p121 (r) Sinclair Stammers/Science Photo Library, p126–127 Robert74/Shutterstock, p126 (t) PHB.cz (Richard Semik)/Shutterstock, p126 (c) Charles D. Winters/Science Photo Library, p126 (b) ushi/Shutterstock, p127 (t) sciencephotos/Alamy Stock Photo, p127 (b) Charles D. Winters/Science Photo Library, p128 (t) Adisa/Shutterstock, p128 (c) Yuri Samsonov/Shutterstock, p128 (b) ipuwadol/Shutterstock, p129 (t) Andrey_Kuzmin/Shutterstock, p129 (b) Andrew Lambert Photography/Science Photo Library, p129 (br) Andrew Lambert Photography/Science Photo Library, p130 (t) Martyn F. Chillmaid/Science Photo Library, p130 (b) Andrew Lambert Photography/Science Photo Library, p131 (t) Martyn F. Chillmaid/Science Photo Library, p131 (b) Heritage Image Partnership Ltd/Alamy Stock Photo, p132 (cl) sciencephotos/Alamy Stock Photo, p132 (cr) sciencephotos/Alamy Stock Photo, p134 NASA/Science Photo Library, p135 (t) Elena Pominova/Shutterstock, p135 (b) Andrew Lambert Photography/Science Photo Library, p136 (t) signature photos/Shutterstock, p136 (c) Hong Vo/Shutterstock, p136 (bl) M. Unal Ozmen/Shutterstock, p136 (br) Steve Stock/Alamy Stock Photo, p137 sciencephotos/Alamy Stock Photo, p138 (tl) chrisbrignell/Shutterstock, p138 (tr) DenisNata/Shutterstock, p138 (b) Birgit Reitz-Hofmann/Shutterstock, p139 (tl) Richard Heyes/Alamy Stock Photo, p139 (tr) whiteboxmedia limited/Alamy Stock Photo, p139 (b) ffolas/Shutterstock, p140 Andrew Lambert Photography/Science Photo Library, p142 Manor Photography/Alamy Stock Photo, p149 Charles D. Winters/Science Photo Library, p150–151 Mark Garlick/Science Photo Library, p150 (t) Orhan Cam/Shutterstock, p150 (c) alice photo/Shutterstock, p150 (b) Triff/Shutterstock, p151 majusko95/Shutterstock, p154 Orhan Cam/Shutterstock, p155 (tl) Joyce Photographics/Science Photo Library, p155 (tr) Stocktrek Images/Getty Images p155 (bl) Twonix Studio/Shutterstock, p155 (br) Herve Conge, ISM/Science Photo Library, p156 (t) Pasieka/Science Photo Library, p156 (b) Leene/Shutterstock, p157 Peter Hulla/Shutterstock, p158 (t) Ivan Varyukhin/Shutterstock, p158 (b) Ollie Taylor/Shutterstock, p159 (t) Lisa S./Shutterstock, p159 (ct) sigur/Shutterstock, p159 (c) Bortel Pavel – Pavelmidi/Shutterstock, p159 (cb) suriya phonlakorn/Shutterstock, p159 (b) LesPalenik/Shutterstock, p161 (t) Martin Bond/Science Photo Library, p161 (b) Tao Chuan Yeh/Staff/AFP/Getty Images p162 (t) Triff/Shutterstock, p162 (b) Yuriy Kulik/Shutterstock, p163 mironov/Shutterstock, p166 majusko95/Shutterstock, p167 David Carillet/Shutterstock, p168 racorn/Shutterstock, p169 tdhster/Shutterstock, p174–175 Dr Jeremy Burgess/Science Photo Library, p174 Science Photo Library/Science Photo Library, p175 My Planet/Alamy Stock Photo, p177 (t) Bon Appetit/Alamy Stock Photo, p177 (c) My Planet/Alamy Stock Photo, p177 (b) MedicalRF.com/Getty Images, p181 (tl) junyanjiang/Shutterstock, p181 (tr) Mr. Suttipon Yakham/Shutterstock, p181 (b) Martyn F. Chillmaid/Science Photo Library, p182 (t) Howard Klaaste/Shutterstock, p182 (b) itsmejust/Shutterstock, p183 Medical Photo NHS Lothian/Science Photo Library, p185 (l) Science Photo Library/Science Photo Library, p185 (r) Geoff Tompkinson/Science Photo Library, p185 (r) Pr Michel Brauner, ISM/Science Photo Library, p186 Astrid & Hanns-Frieder Michler/Science Photo Library, p189 Steve Gschmeissner/Science Photo Library, p190 Kateryna Kessariiska/Shutterstock, p191 (t) Microscape/Science Photo Library, p191 (c) Kevin & Betty Collins, Visuals Unlimited/Science Photo Library, p191 (b) Jezper/Shutterstock, p192 USBFCO/Shutterstock, p193 USBFCO/Shutterstock, p197 (l) Edward Kinsman/Science Photo Library, p197 (r) Dr David Furness, Keele University/Science Photo Library, p198–199 Whatafoto/Shutterstock, p198 (t) Pavel Bredikhin/Shutterstock, p198 (b) Wildlife GmbH/Alamy Stock Photo, p199 Kiorio/Shutterstock, p200 Pixeljoy/Shutterstock, p202 (t) Jaubert Images/Alamy Stock Photo, p202 (c) Cal Vornberger/Alamy Stock Photo, p203 Kodda/Shutterstock, p204 (t) Len Wilcox/Alamy Stock Photo, p204 (b) Dennis Cox/Alamy Stock Photo, p205 Carlos Amarillo/Shutterstock, p206 (t) Images of Africa Photobank/Alamy Stock Photo, p206 (b) Mark Beckwith/Shutterstock, p207 dabjola/Shutterstock, p208 Ian Gowland/Science Photo Library, p209 (t) Tim Gainey/Alamy Stock Photo, p209 (bl) Pavel Bredikhin/Shutterstock, p209 (br) Wildlife GmbH/Alamy Stock Photo, p210 Norman Chan/Shutterstock, p212 (t) Serge Vero/Shutterstock, p212 (c) Vlada Z/Shutterstock, p212 (bl) Dario Lo Presti/Shutterstock, p212 (br) spaxiax/Shutterstock, p213 (l) D. Kucharski K. Kucharska/Shutterstock, p213 (cl) Kiorio/Shutterstock, p213 (cr) Charles E Mohr/Science Photo Library, p213 (r) Gilles Mermet/Science Photo Library, p214 (t) Lu Mikhaylova/Shutterstock, p214 (c) Dr. John Brackenbury/Science Photo Library, p214 (b) amnachphoto/Shutterstock, p220–221 Jolanta Wojcicka/Shutterstock, p220 (t) Khoroshunova Olga/Shutterstock, p220 (c) Chris Humphries/Shutterstock, p220 (b) Eye of Science/Science Photo Library, p221 (t) Sue McDonald/Shutterstock, p221 (b) wong yu liang/Shutterstock, p222 (t) Sergey Novikov/Shutterstock, p222 (cl) Eric Isselee/Shutterstock, p222 (cr) cynoclub/Shutterstock, p222 (b) Erik Lam/Shutterstock, p224 (t) Karen McGaul/Shutterstock, p224 (b) EpicStockMedia/Shutterstock, p225 (t) wavebreakmedia/Shutterstock, p225 (b) Romeo Gacad/Staff/AFP/Getty Images p226 (t) Linda Bucklin/Shutterstock, p226 (ct) Hemis/Alamy Stock Photo, p226 (cb) Menno Schaefer/Shutterstock, p226 (b) Constant/Shutterstock, p227 (t) Steve McWilliam/Shutterstock, p227 (b) Martin Fowler/Shutterstock, p231 (t) Eye of Science/Science Photo Library, p231 (b) Eye of Science/Science Photo Library, p232 Juergen Berger/Science Photo Library, p233 (l) BlueRingMedia/Shutterstock, p233 (r) 3d4medical.com/Science Photo Library, p234 P. Saada/Eurelios/Science Photo Library, p235 herjua/Shutterstock, p236 (t) wong yu liang/Shutterstock, p236 (c) RetroClipArt/Shutterstock, p236 (b) RetroClipArt/Shutterstock, p237 malost/Shutterstock.